Mathem

Reasoning™
Level A

Developing Math & Thinking Skills

Series Titles
Mathematical Reasoning™ Beginning
Mathematical Reasoning™ Level A
Mathematical Reasoning™ Level B
Mathematical Reasoning™ Level C
Mathematical Reasoning™ Book 1
Mathematical Reasoning™ Book 2

Written by
Linda Brumbaugh
Doug Brumbaugh

Graphic Design by
Karla Garrett
Linda Laverty
Anna Allshouse
Doug Brumbaugh

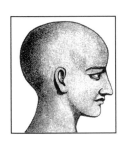

© 2008, 2006
THE CRITICAL THINKING CO.™
(BRIGHT MINDS™)
Phone: 800-458-4849 Fax: 831-393-3277
www.CriticalThinking.com
P.O. Box 1610 • Seaside • CA 93955-1610
ISBN 978-1-60144-181-2

Printed in the USA

About the Authors

Linda S. Brumbaugh

I retired after teaching a total of 31 years in grades three, four, and five. Both my BS from the University of Florida and Masters from the University of Central Florida are in Elementary Education. As I look back over my teaching career, I enjoyed seeing the excitement on the children's faces as they encountered new concepts, worked with a manipulative, experienced some new mathematical application, or played a new mathematical game. It was stimulating when they solved an intricate problem, discovered something new to them, or got caught up in some new mathematical trick. As they got excited about learning, so did I. Each day of every year brought some new learning opportunity for me and for the children. I continue to work with pre-school and elementary age children in the Sunday school system of our church. Our intent is to convey some of that excitement to each child who uses this book.

Douglas K. Brumbaugh

Depending on how you count, I have been teaching over 50 years. I taught in college, in-service, or K-12 almost daily. I received my BS from Adrian College and Masters and Doctorate in Mathematics Education from the University of Georgia. Students change, classroom environments change, the curriculum changes, and I change. The thoughts and examples used here are based on my teaching experiences over the years. The pages in this book are designed to spark the interest of each child who works with them.

TABLE OF CONTENTS

NCTM STANDARDS	Number and Operations	Algebra	Geometry	Measurement	Data Analysis and Probability
SKILLS Addition	13, 14, 41, 44, 45, 47, 60, 68, 69, 72, 73, 76, 77, 79, 80, 82, 85, 88, 89, 132, 141, 160, 163, 164, 165, 166, 167, 168, 174, 181, 187, 192, 199, 202, 206, 207, 209, 213, 218, 226	39, 59, 62, 70, 71, 81, 86, 90, 182, 184, 216, 219			
Bar Graph					100, 118, 120, 121, 124, 125, 178, 179, 203, 205
Calendar				173	
Capacity				225, 247	
Coins				135, 136, 177, 214, 247	
Count	11, 24, 33, 34, 37, 38, 41, 42, 43, 44, 45, 46, 47, 52, 63, 72, 73, 76, 77, 78, 79, 80, 110, 111, 131, 138, 139, 140, 141, 144, 145, 146, 153, 154, 159, 160, 163, 164, 165, 166, 167, 168, 171, 180, 181, 183, 187, 188, 189, 192, 194, 199, 204, 206, 207, 212, 215, 226, 230, 231	8, 9, 10, 39, 65, 70, 71, 81, 91, 92, 95, 96, 97, 98, 122, 143, 172, 182, 184, 185, 186, 227, 228, 229	84	22, 135	75, 118, 120, 121, 124, 125, 178, 179, 203
Fractions	128, 132, 133, 162, 175, 176, 197		222	129, 130, 196	
Language	21, 35, 36, 38, 40, 41, 43, 55, 61, 88, 93, 94, 106, 110, 111, 131, 133, 137, 161, 187, 201, 233	15, 16, 29, 57, 65, 86, 90, 91, 92, 95, 96, 97, 98, 116, 119, 186, 216, 219, 228, 229, 242, 244, 246	48, 51, 58, 74, 83, 155, 157, 190, 191, 210, 221, 222, 237	135, 149, 150, 151, 152, 156, 158, 173, 177, 195, 200, 214, 241, 245, 247	67, 75, 87, 100, 113, 118, 120, 123, 124, 125, 126, 127, 142, 178, 179, 203, 217, 223, 232, 234
Length		53, 103		156, 158	99
Likelihood				149, 152, 173	67, 123, 126, 127, 142, 232, 234

TABLE OF CONTENTS (Cont.)

How to Use this Book

The skills and concepts in this book spiral throughout the book. That means that you will see a topic dealt with for a few pages and then a gap before it is covered again. We do that so your child has some time to develop and mature before dealing with more complex aspects of the skill/concept.

Our suggestion is that you proceed through the book page by page. However, if your child is interested in a given topic and seems to want more, it would not be unreasonable for you to skip to the next level of that topic and do more of it.

Warning: Most 3 and 4 years old can be taught to add, subtract, and reason mathematically, but just as all children do not grow at the same rate, not all brains develop at the same rate. If your child struggles, don't be alarmed and jump to conclusions about your child's intelligence. Children's brains develop at different rates—especially young children.

Teaching Suggestions

Important: Keep learning fun and avoid frustrating your child. Work around your child's attention span. As a adult, you have a great advantage because most young children love to spend time with "big" people. If you keep learning fun, you will have an energetic pupil who looks forward to each lesson.

There is no one correct way to teach the skills presented in this book. Have fun figuring out different ways to relate the skills to your child's daily life.

Using Concrete Objects: If your child struggles with an activity, you can come back to it later or try recreating the activity using objects your child can touch and feel such as counting bears, walk on number lines, or base 10 blocks. These hands-on supplements can also help build a background for future work.

Walk on Number Lines:

You will see several pages that involve number lines. Younger children can accomplish solid formative development while using a "walk on number line." For this, you would put footprints, squares, or colored pieces of paper on the floor at a convenient distance apart. Then your child would step from one to the next, counting the steps taken.

This provides several background bits of information:

- The space between points is what is being counted.
- When counting, you do not count the point where you start.
- You can count some steps and then count some more (like 2 + 3) to get a total number of steps.

- You can start in one direction and then go back the other way (leading to subtraction).

Base 10 Blocks

Base 10 blocks can be used to help children learn addition facts, particularly when the sum is greater than 9. At this level, there are 2 basic pieces to be used:

The unit, ⬚ , and the long, ▭▭▭▭▭▭▭▭▭▭ .

Representatives of these pieces can be cut from construction paper or cardboard. It is important that the long be ten times the length of the side of the unit. That is, if the unit is a square that is one inch on a side, then the long would be a rectangle that is one inch wide and 10 inches long.

Initially your child should become familiar with making trades; 10 units would trade for one long (or 10 ones would trade for one ten), and vice versa. These two relationships are essential background for future regrouping in addition and subtraction.

Once trades are made with ease, and even before, the units can be used as counters to work out problems like 2 + 3, 4 + 1, 2 + 7, and so on. At first, keep the sums small, but as your child becomes more familiar with larger numbers, increase the sum. In the beginning, sums should be 9 or less because no regrouping is involved. If your child is ready for a challenge, use more than two addends (1 + 2 + 3), still limiting the sum to 9 or less.

Eventually the sums will be 10 or larger and that is when the trading exercises begin to pay dividends. Your child should reflexively want to trade 10 ones for 1 ten so that a sum like 7 + 5 would be shown as 10 + 2, which is interpreted as one ten (or long) and two ones (or units). Still, 10 + 2 is a common way of writing values and is called "expanded notation." This too is something your child will encounter in the early school years.

Virtual Manipulatives:

You might be interested in checking out the National Library of Virtual Manipulatives at http://nlvm.usu.edu/en/nav/vlibrary.html. Click on "Virtual Library" at the top of the page. Then click on the Number and Operations row and the PreK-2 column. If you click on the second entry, "Base Blocks," you can create your own problem. To see how it works, click on the unit cube and drag it to the space below it. Do that at least 12 times. Then, using your mouse, draw a rectangle around 10 of the pieces you see in the workspace. You will see those 10 units trade for one ten. That is a powerful image for a child.

Please note that there are base blocks for addition and subtraction on that same Number and Operations row and Pre K-2 cell on the table.

The activities in this book are written to the standards of the National Council of Teachers of Mathematics (NCTM).

Touch each shape and say its name.

Triangles

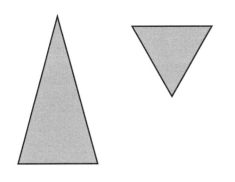

Circles

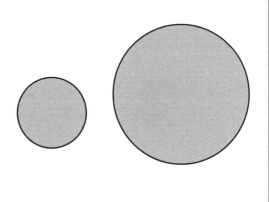

Squares

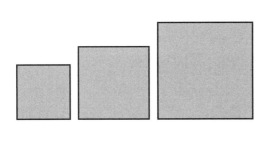

Rectangles

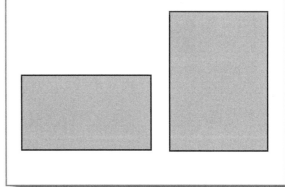

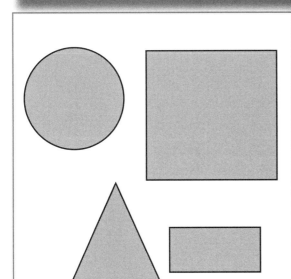

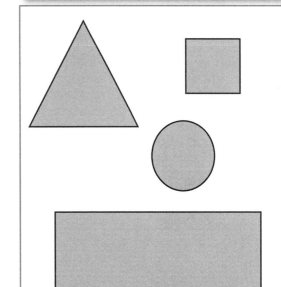

Draw line segments to connect like colors.

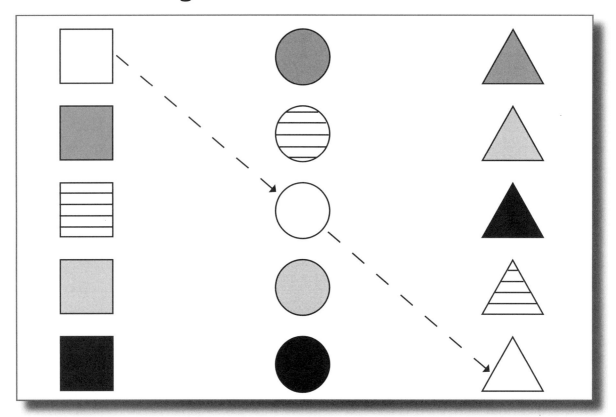

Draw line segments to connect like shapes.

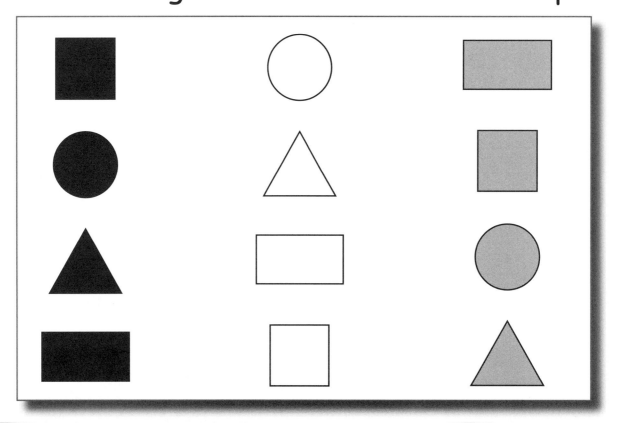

How Many?

Trace by following the arrows.

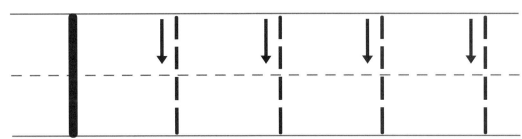

How Many?

Trace by following the arrows.

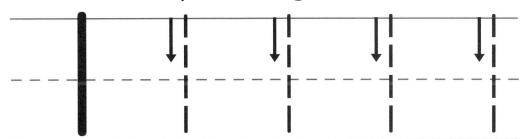

How Many?

Trace by following the arrows.

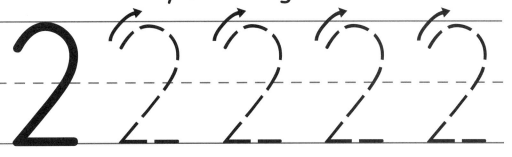

How Many?

Trace by following the arrows.

How Many?

Trace by following the arrows.

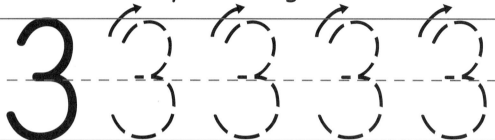

How Many?

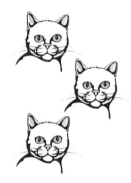

Trace by following the arrows.

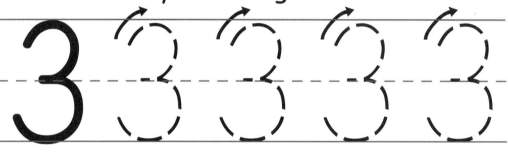

How Many?

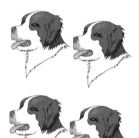

Trace by following the arrows.

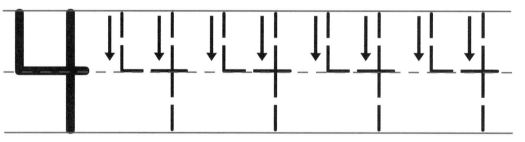

How Many?

Trace by following the arrows.

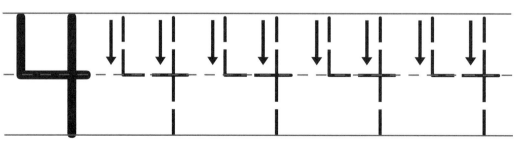

How Many?

Trace by following the arrows.

5 5 5 5 5

How Many?

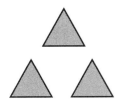

Trace by following the arrows.

5 5 5 5 5

How Many?

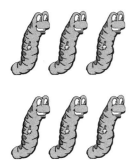

Trace by following the arrows.

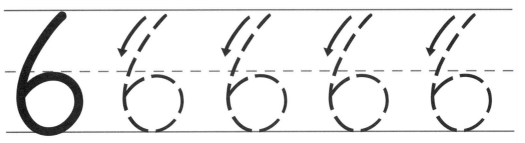

How Many?

Trace by following the arrows.

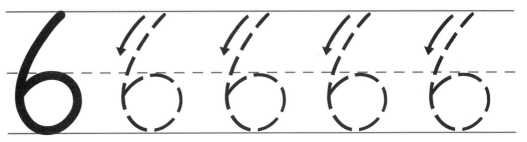

How Many?

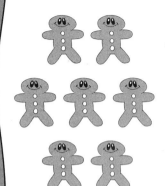

Trace by following the arrows.

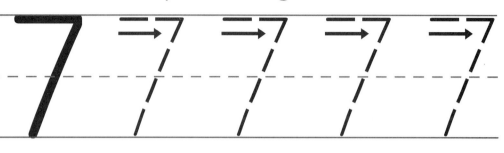

How Many?

Trace by following the arrows.

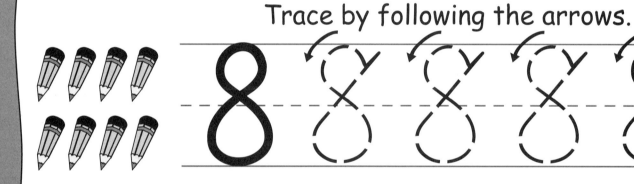

How Many?

Trace by following the arrows.

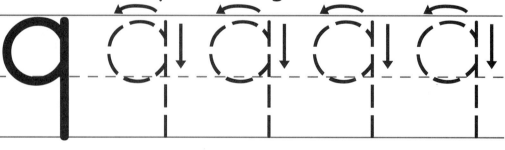

Trace each number, then draw a line segment to the matching picture.

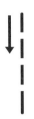

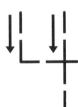

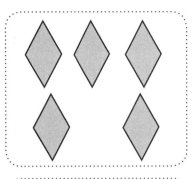

Trace each numeral, then circle the correct amount.

Butterflies

1   2   3

Dog

1   2   3

Chickens

1   2   3

Trace each numeral, then circle the correct amount.

Snails

4   5   6

Clowns

4   5   6

Pigs

4   5   6

Trace each numeral, then circle the correct amount.

Cats

7   8   9

Cows

7   8   9

Horses

7   8   9

Trace each numeral, then circle the number of stars.

Circle each set with the <u>same</u> number of objects.

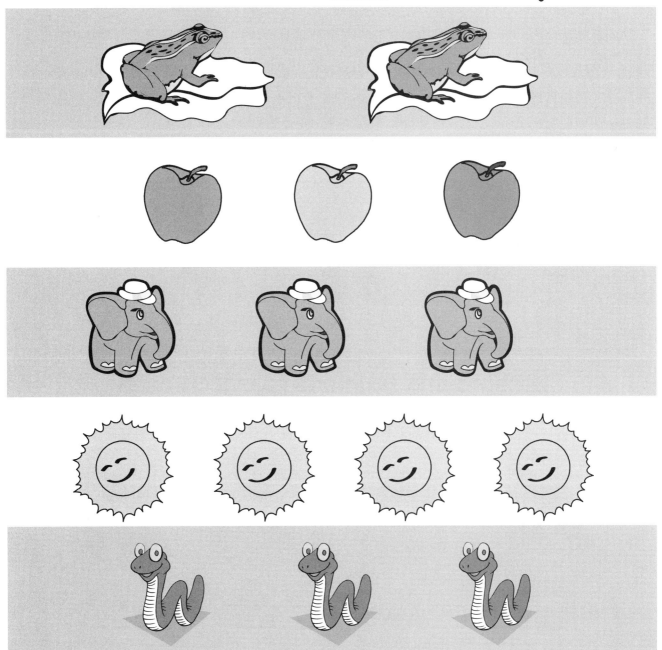

Draw a line segment to match each bat and ball of the same color.

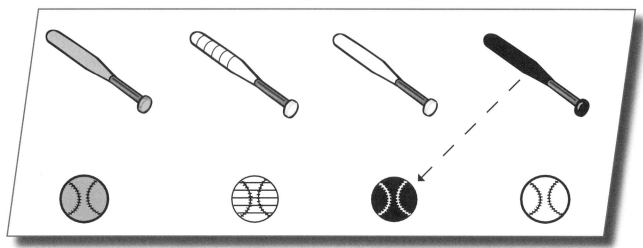

Draw line segments to match the rows of balls and jacks that have the same number.

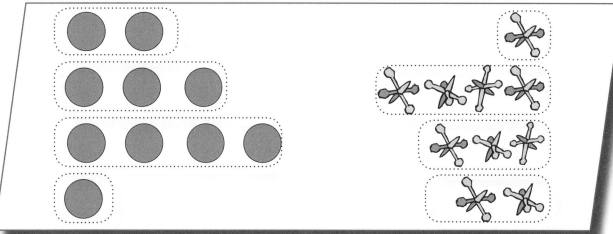

Draw a line segment to match each square with its missing piece.

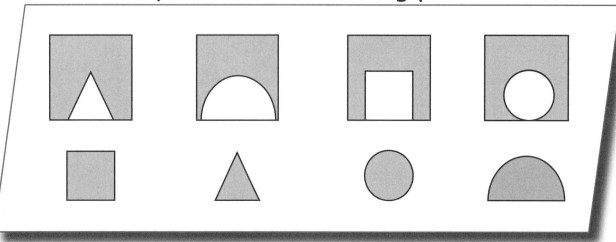

Match pairs of dominos to make 3.

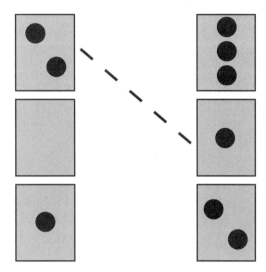

Match pairs of dominos to make 4.

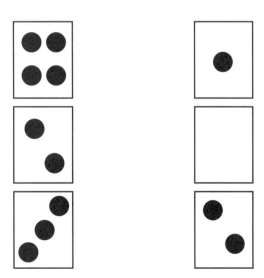

Match pairs of dominos to make 5.

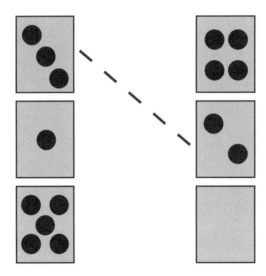

Match pairs of dominos to make 6.

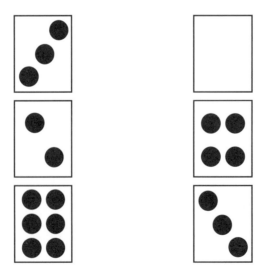

Same, Fewer, or More

The black hat has the same, fewer, or more flowers.

The gray hen has the same, fewer, or more eggs.

The round fish bowl has the same, fewer, or more goldfish.

The stars have the same, fewer, or more objects than the other set.

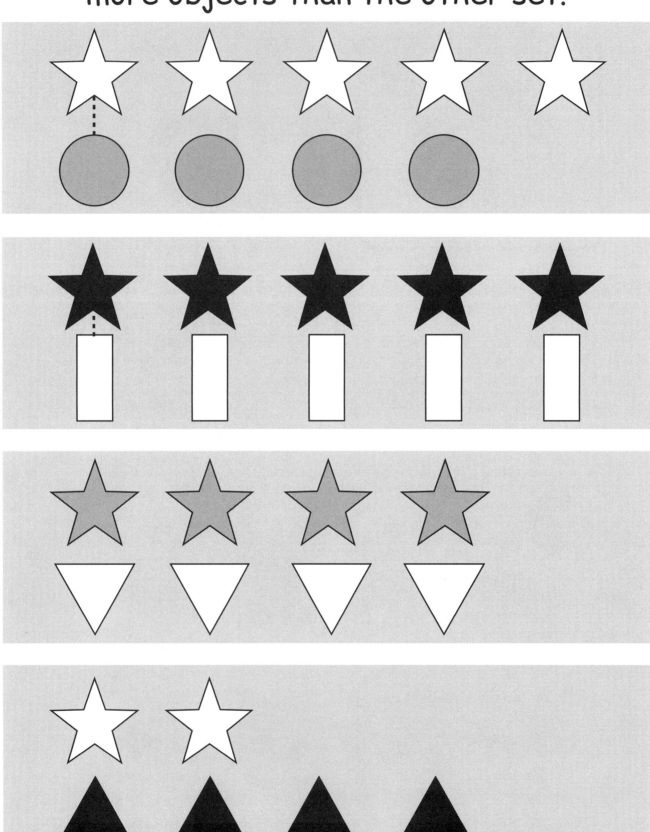

Even Numbers

When every face in a set has a partner, the number of faces in the set is an even number. Draw line segments to try to connect each face with a partner. Circle or say whether the set is even.

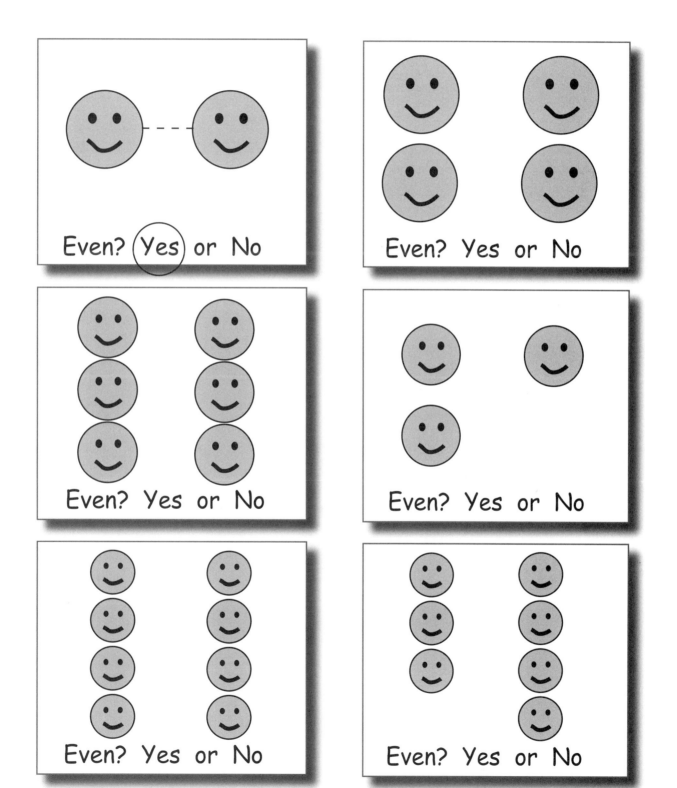

Draw line segments to connect each dog to the bones of the same color.

Draw line segments to connect each monkey to the bananas of the same color.

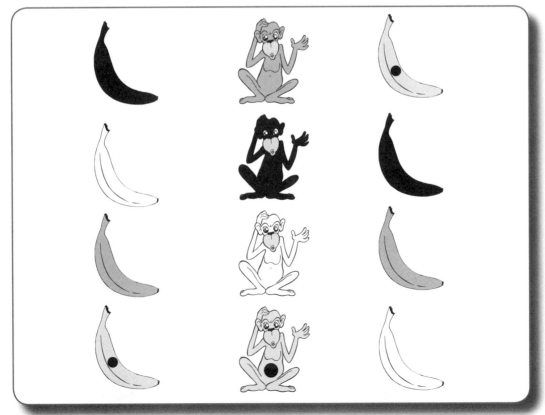

Even or Odd
When a face in a set does not have a partner, the number of faces in the set is an odd number. Draw line segments to try to connect each face with a partner. Circle or say whether the set is even or odd.

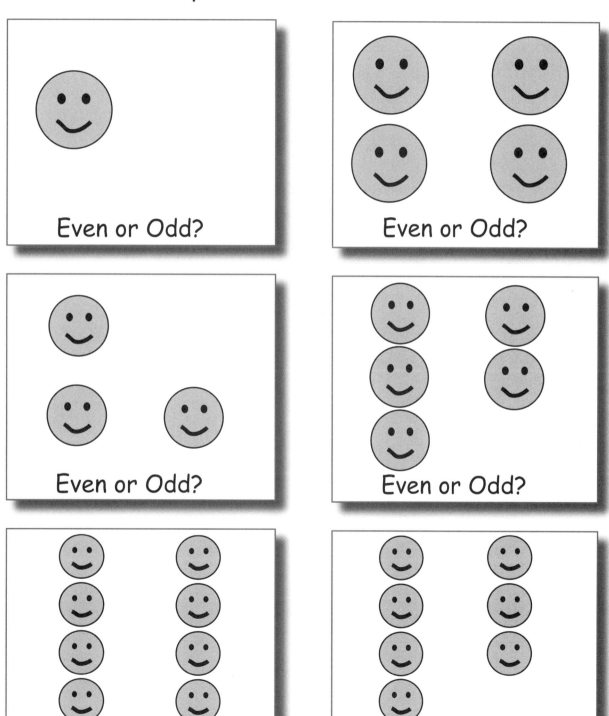

Even or Odd?

Even or Odd?

Even or Odd?

Even or Odd?

Even or Odd?

Even or Odd?

Even or Odd
Trace each numeral, then circle the correct amount. Draw line segments to connect the partners to find out if each number is even or odd.

7   8   9   Even or Odd?

1   2   3   Even or Odd?

4   5   6   Even or Odd?

Use a crayon to make each set an even number.

Draw a line from each object to the picture where it belongs. Use a different color crayon for each picture. Accept any answer that is rationally justified.

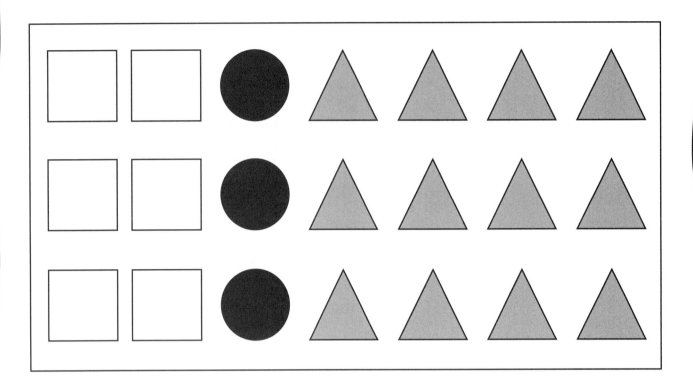

Color with a crayon to make each sentence true.

1. There are more ☐ than _____.

2. There are fewer ☐ than _____.

3. There are fewer ◯ than _____.

4. There are more ☐ and _____ than _____.

Even or Odd
Trace each numeral, then circle the correct answers.

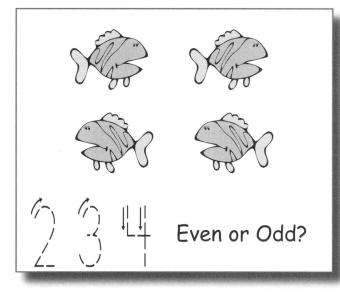

2 3 4 Even or Odd?

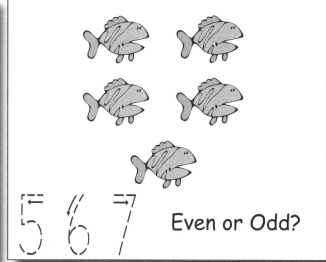

5 6 7 Even or Odd?

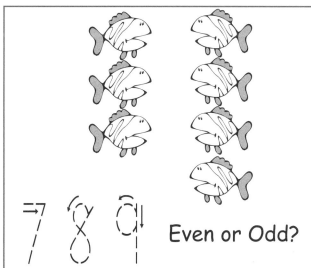

7 8 9 Even or Odd?

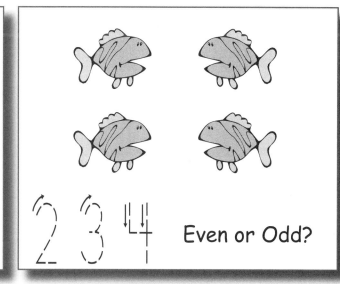

2 3 4 Even or Odd?

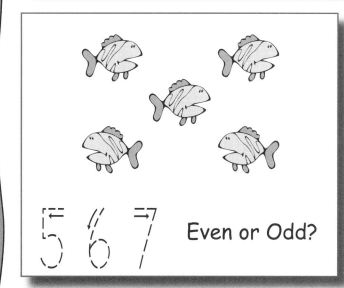

5 6 7 Even or Odd?

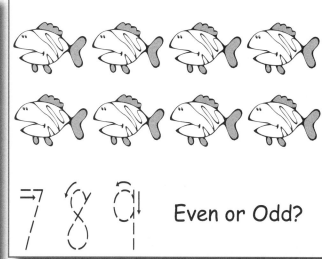

7 8 9 Even or Odd?

Trace each numeral, then circle the correct amount.

 Count stars.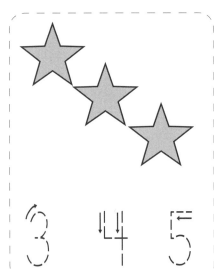

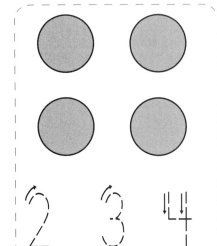 Count circles.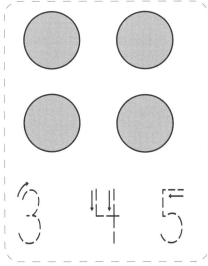

Use a line segment to connect the matching sets.

Even or Odd

Trace each numeral, then circle the correct answers.

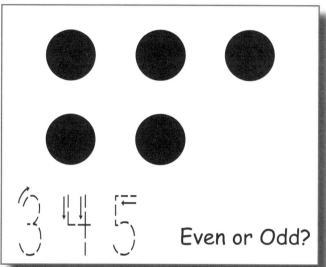

Even or Odd?

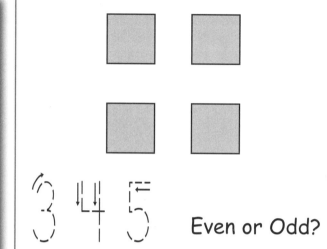

Even or Odd?

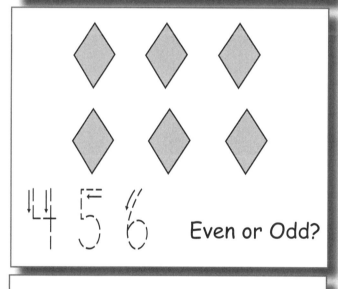

Even or Odd?

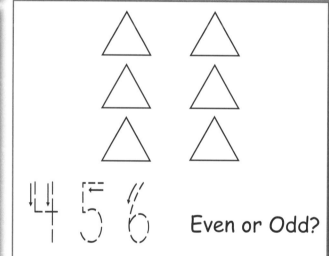

Even or Odd?

Even or Odd?

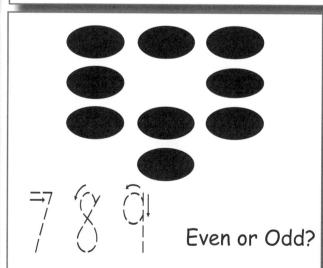

Even or Odd?

Ordinal Numbers

Which fish will probably get the worm: first? second? third?

1. Point to the first bird in line.
2. Point to the second bird in line.
3. Point to the third bird in line.

Ordinal Numbers

The animals are lined up to dive into the pool. Point to the divers in order: <u>first</u>, <u>second</u>, <u>third</u>, <u>fourth</u>, <u>fifth</u>.

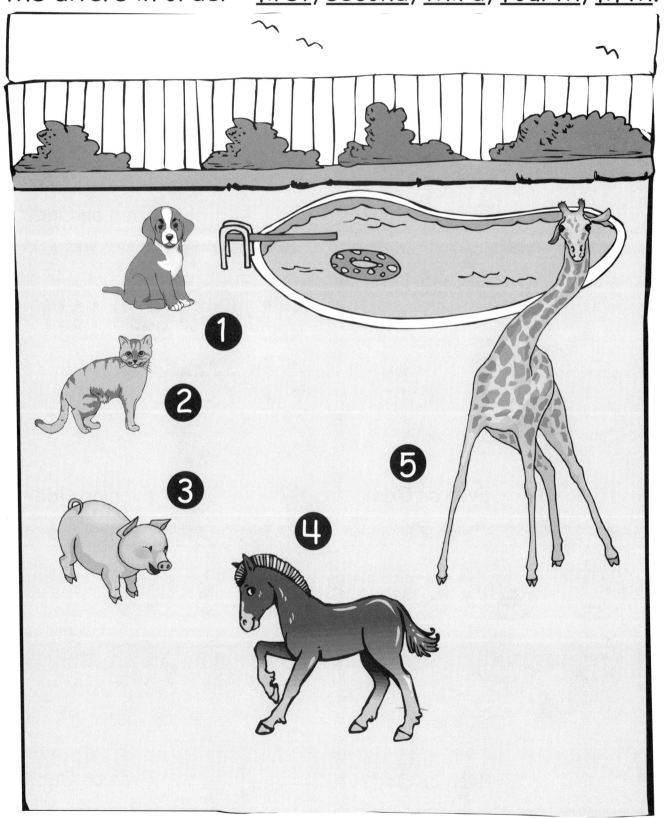

Even or Odd

Trace each numeral, then circle the correct answers.

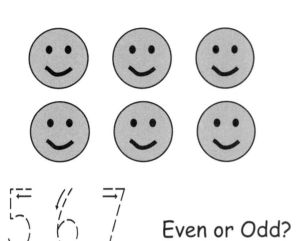

Even or Odd?

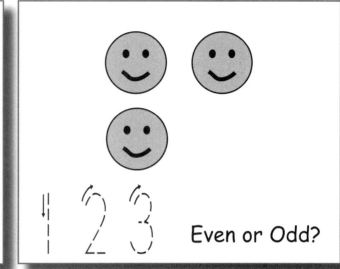

Even or Odd?

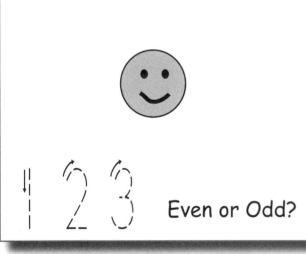

Even or Odd?

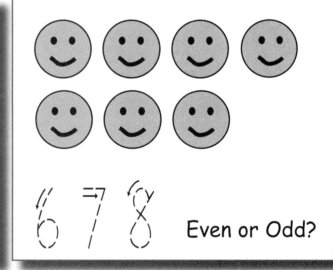

Even or Odd?

Even or Odd?

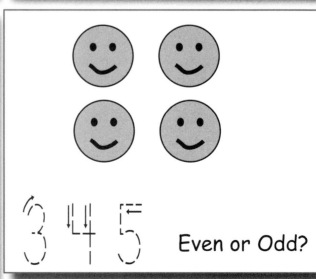

Even or Odd?

Point or say the answer.

1. Point to the first car in line.

2. Point to the fifth car in line.

3. Point to the second car in line.

4. Point to the car in the middle of the line.

5. Point to the car that is next to the last in line.

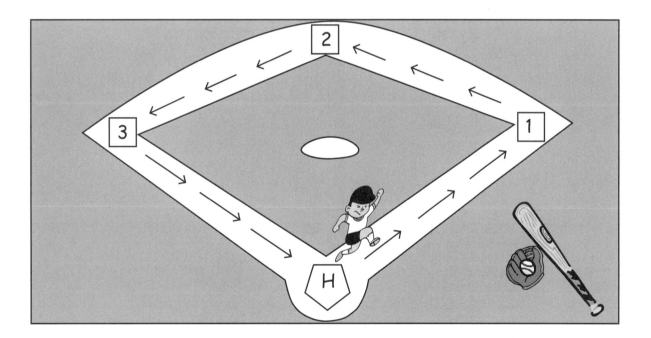

1. Which base will the runner touch second?

2. Which base will the runner touch last?

3. Which base will the runner touch first?

4. Which base will the runner touch third?

Even or Odd
Trace each numeral, then circle the correct answers.

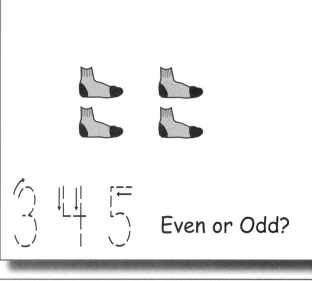

Even or Odd?

Even or Odd?

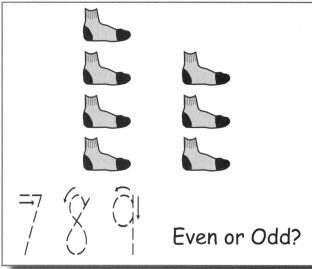

Even or Odd?

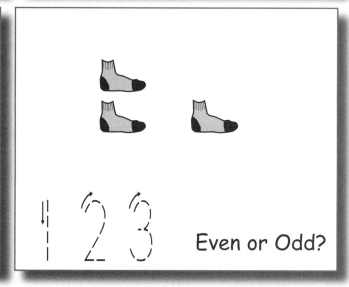

Even or Odd?

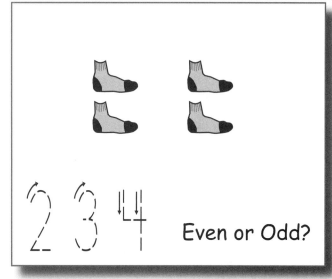

Even or Odd?

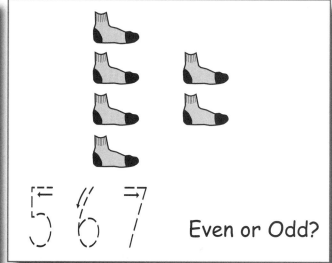

Even or Odd?

Use a line segment to match the sets with the same numbers.

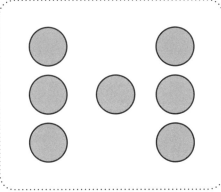

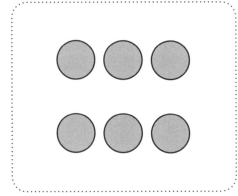

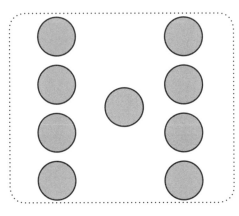

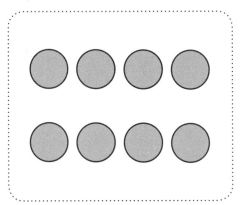

Even or Odd

Trace each numeral, then circle the correct answers.

Even or Odd?

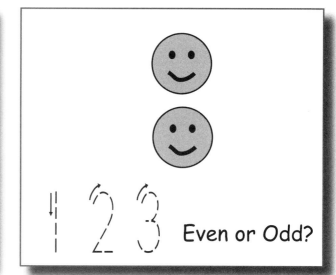

Even or Odd?

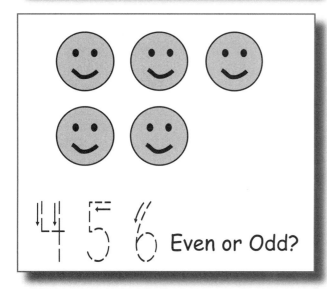

Even or Odd?

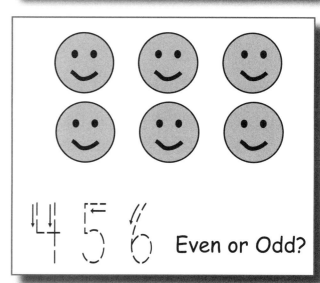

Even or Odd?

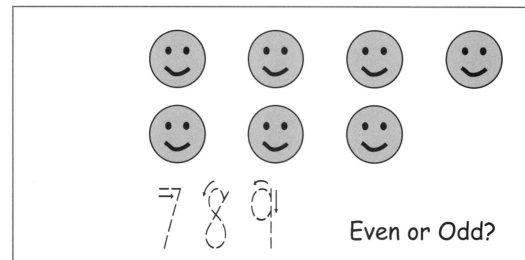

Even or Odd?

Draw line segments to connect the sets.

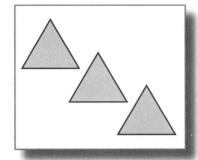 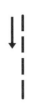 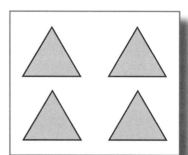

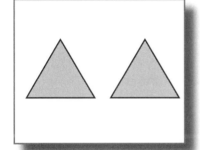  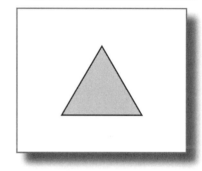

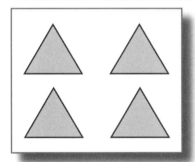  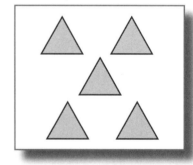

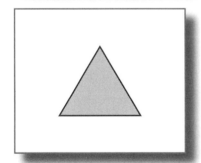 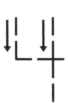 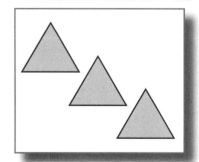

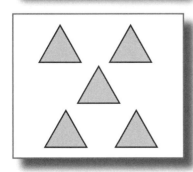  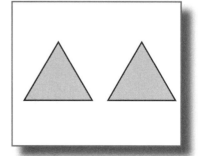

Draw line segments to connect the sets.

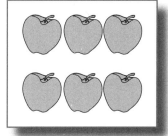

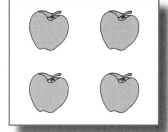

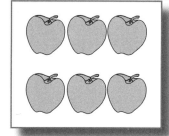

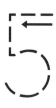

Circle the correct animal.

First?

Fourth?

Third?

Second?

Fifth?

The fish are racing to get the worm.

Which fish is <u>second</u>?

Which fish is <u>third</u>?

Which fish is <u>first</u>?

Which fish is <u>fifth</u>?

Which fish is <u>fourth</u>?

Trace each number, then draw a line segment to its matching picture.

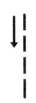

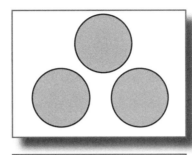

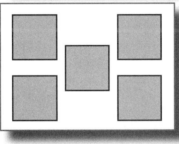

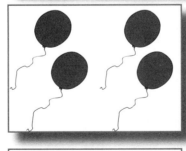
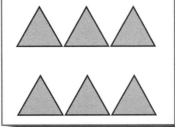

Point or say the answer to each question.

A little frog sat on a lily pad.
A spider was spinning his web.
A little fly came buzzing by.
Out came a long, pink tongue.
No more fly!

1. How many animals are in the picture?
(Insects are animals.)

2. How would the picture change at the end of the story? (Fly is gone.)

3. Which animal in the picture is the biggest?

4. Which animal in the picture is the smallest?

5. How many animals in the picture can swim?

6. How many animals in the picture can fly?

7. Why does a spider spin a web?

*Accept any answer that can be logically justified by the child.

Use the clues to draw a line segment to match each child with the correct number of fish. Write the number of fish in each group.

Lee's fish have dots on them.

_____

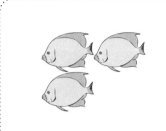

Beth has the most fish.

_____

Maria has two more fish than Lee.

_____

Tyler has one less than Maria.

_____

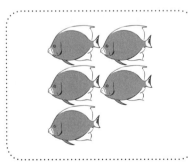

Numeral Guide

The <u>white</u> horse is <u>first</u>.

1. Circle the <u>second</u> horse.

2. Put a square around the <u>third</u> horse.

3. Put a triangle around the <u>fourth</u> horse.

4. Put a rectangle around the <u>fifth</u> horse.

This **+** sign means to add the number of things in each set. This **=** sign means to write the sum of the number of things in each set.

Count the bugs in each set and write the sum.

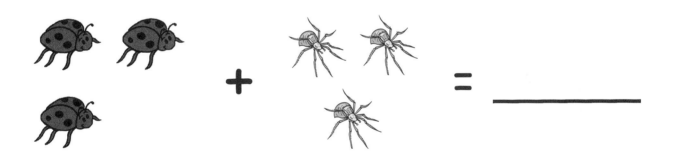

Numeral Guide

1 2 3 4 5 6 7 8 9

Trace each numeral, then circle the number of stars.

Circle each set with the <u>same</u> number of objects.

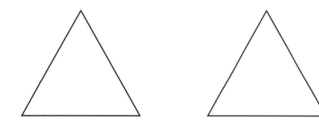 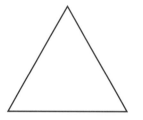

Zero Means None

There are <u>no</u> fish in the bowl.
There are <u>zero</u> (0) fish.

Trace each numeral, then circle
the correct amount.

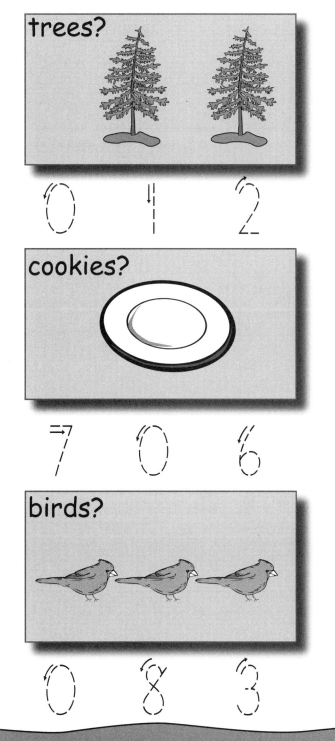

trees?

0   1   2

cookies?

7   0   6

birds?

0   8   3

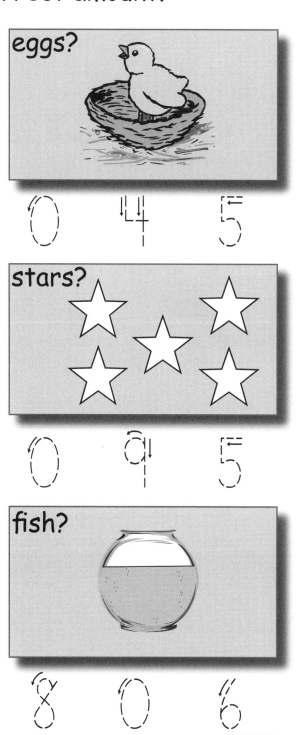

eggs?

0   4   5

stars?

0   9   5

fish?

8   0   6

Count the cubes in each set and say or write how many cubes there are all together.

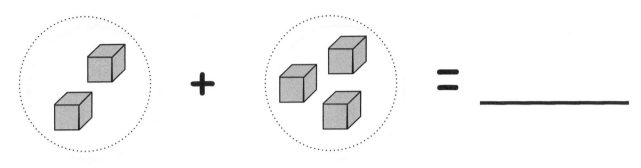

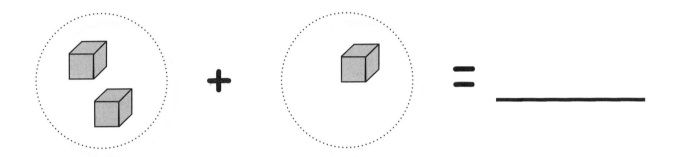

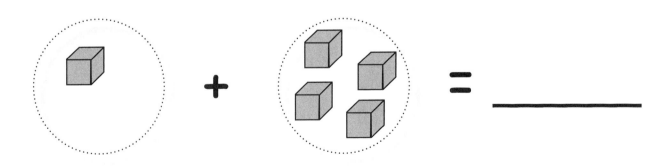

Numeral Guide

1 2 3 4 5 6 7 8 9

See Teaching Suggestion regarding base 10 blocks.

Count the cubes in each set and say or write the sum (how many all together).

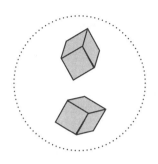 + 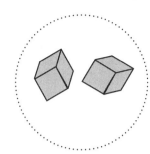 = _____

 + 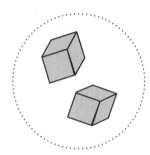 = _____

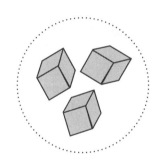 + 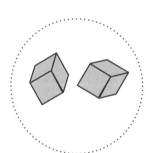 = _____

 + 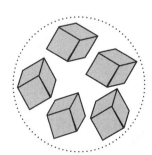 = _____

Numeral Guide

1 2 3 4 5 6 7 8 9

See Teaching Suggestion regarding base 10 blocks.

Trace each number, then draw a line segment to the matching group.

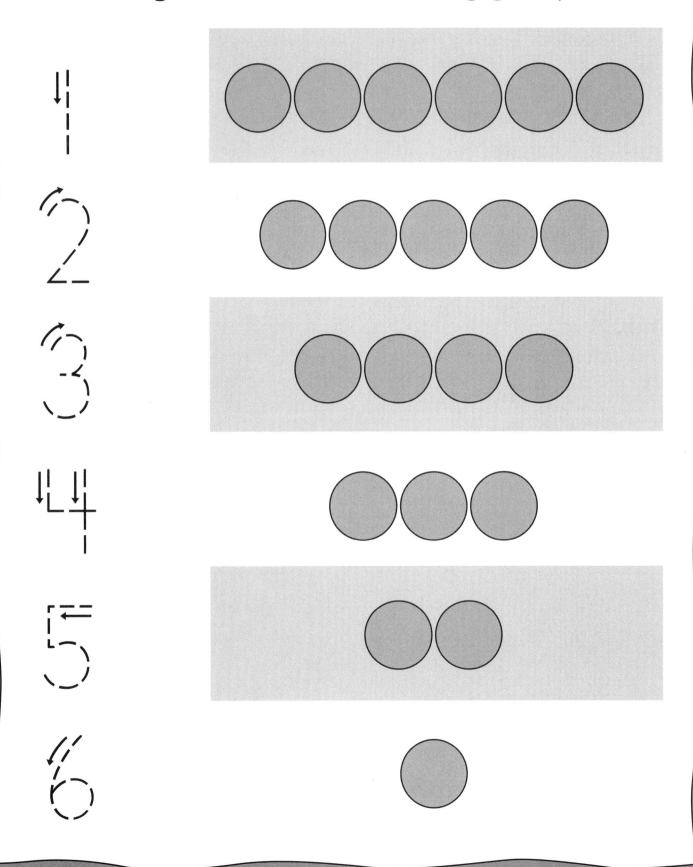

Count the cubes in each set and say or write the sum (how many all together).

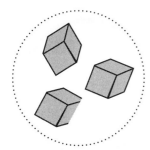 + 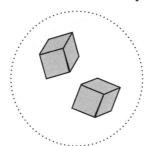 = _____

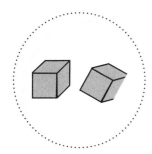 + 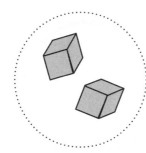 = _____

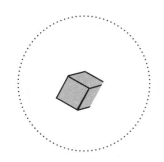 + 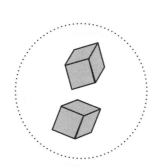 = _____

 + 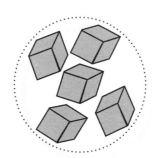 = _____

Numeral Guide

1 2 3 4 5 6 7 8 9

See Teaching Suggestion regarding base 10 blocks.

Draw a square inside all of the triangles.

Draw a triangle inside all of the circles.

Draw a circle inside all of the squares.

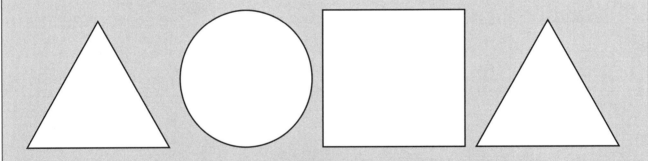

Draw a triangle inside all of the shapes that have corners.

Say, draw, or point to the object that would continue the pattern.

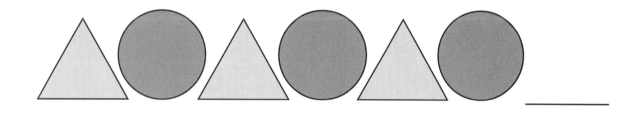

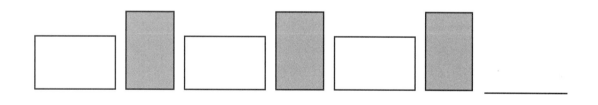

Say, draw, or point to the object that would continue the pattern.

_____

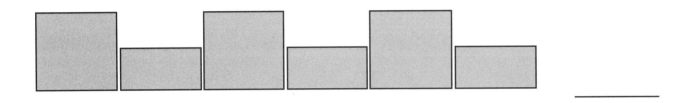

_____

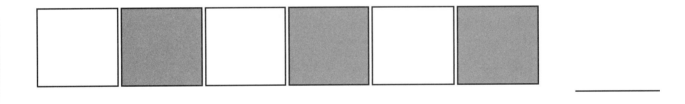

_____

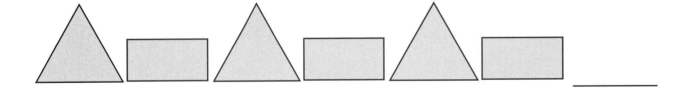

_____

| 1 | 2 | 3 | 4 | 5 | 6 |

_____

Point to each shape and say its name, then follow the directions.

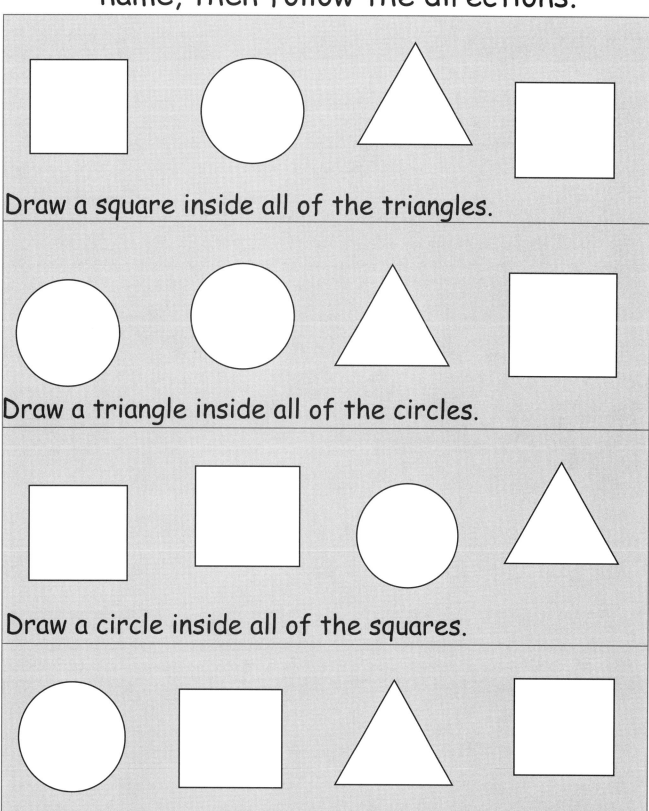

Draw a square inside all of the triangles.

Draw a triangle inside all of the circles.

Draw a circle inside all of the squares.

Draw a triangle inside all of the shapes that have corners.

Trace each number, write each number, and then circle the matching number of triangles in the row.

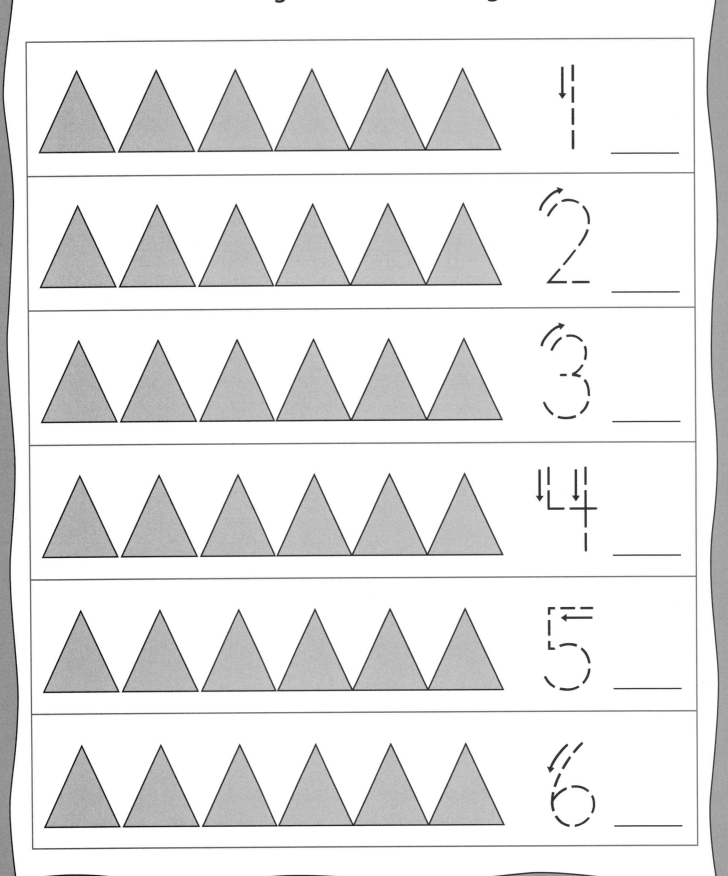

This is a model of a number line.

A given number would be located on this number line by pointing to it, or by tracing the path or distance from zero to the selected number. For example, to locate 3 on the number line, start at 0 and hop to 3, as shown.

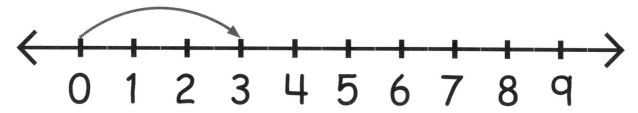

Locate 5 on the number line by tracing the path from 0 to 5.

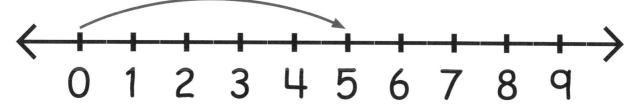

Locate 8 on the number line by tracing the path from 0 to 8.

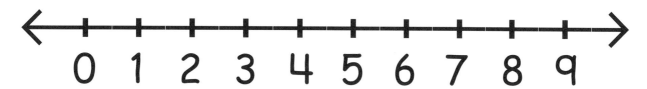

Locate 3 on the number line by tracing the path from 0 to 3.

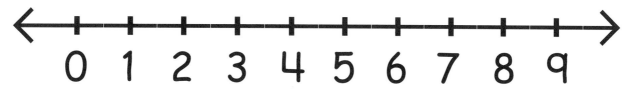

Locate 2 on the number line by tracing the path from 0 to 2.

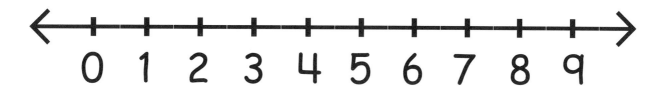

See Teaching Suggestion regarding number lines.

Locate 1 on the number line by tracing the path from 0 to 1.

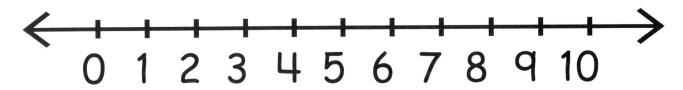

Locate 4 on the number line by tracing the path from 0 to 4.

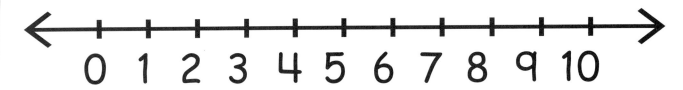

Locate 3 on the number line by tracing the path from 0 to 3.

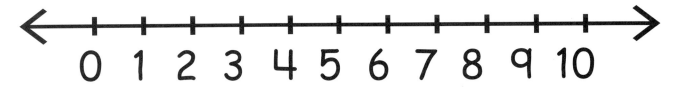

Locate 2 on the number line by tracing the path from 0 to 2.

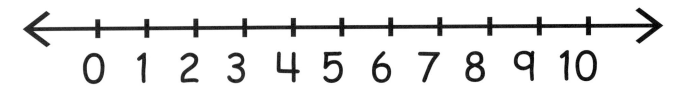

Locate 5 on the number line by tracing the path from 0 to 5.

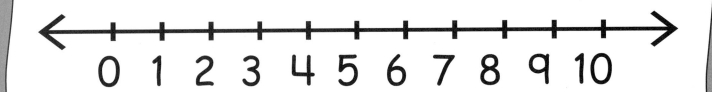

Which shirt has **<u>fewer</u>** stripes?

Which clown has **<u>fewer</u>** balloons?

Which flower has **<u>fewer</u>** bees?

Which dog has **<u>fewer</u>** spots?

Locate 5 on the number line by tracing the path from 0 to 5.

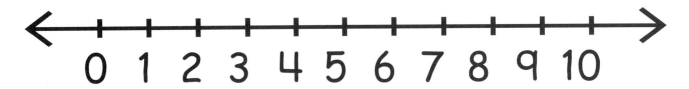

Locate 2 on the number line by tracing the path from 0 to 2.

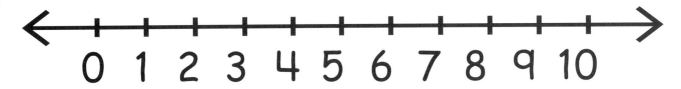

Locate 3 on the number line by tracing the path from 0 to 3.

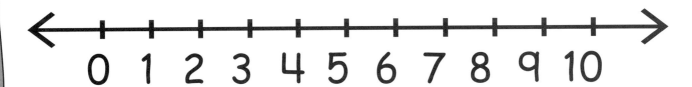

Locate 1 on the number line by tracing the path from 0 to 1.

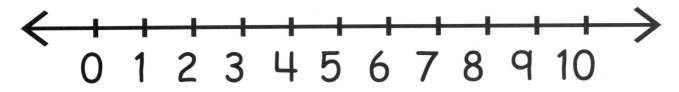

Locate 4 on the number line by tracing the path from 0 to 4.

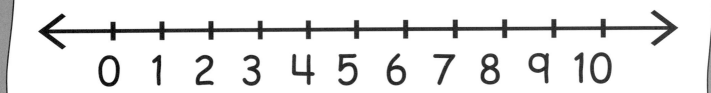

MIND BENDERS®

DIRECTIONS: Fill in the chart using Y for yes or N for no as you solve the puzzle.

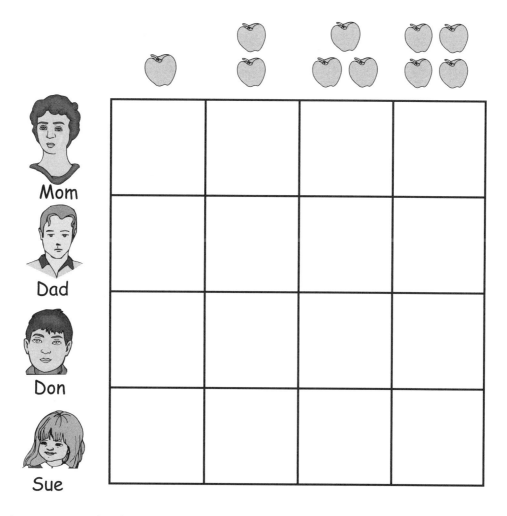

Sue, Don, and their mom and dad all picked apples from their apple tree. Use the clues and chart to find how many apples each person picked.

1. The mom picked the second most apples.
2. Sue picked one less apple than her brother.

* For more activities like this, please see our *Mind Benders®* series.

PANDA BEAR

I am always black and white. You
can see me at the zoo.

Draw the missing parts of the picture, then color
the picture. Can you draw a tree for me to climb?
Can you add something else to the picture?

*For more activities like this, please see our *Thinker Doodles*™ Half and Half Animals series.

The number line is often used to show addition. Starting at zero, hop to the first number. Then, using that number as a starting point, hop the number of steps indicated by the second number. The landing point of the second hop shows the sum of the two numbers. For example, 2 + 3 would appear as:

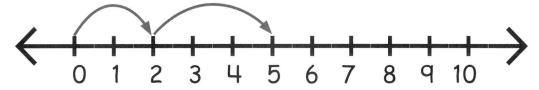

Use the number line to show 1 + 2, then write the sum.

1 + 2= _____

Use the number line to show 2 + 1, then write the sum.

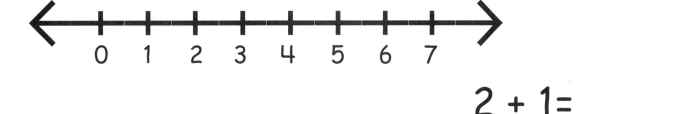

2 + 1= _____

Use the number line to show 2 + 2, then write the sum.

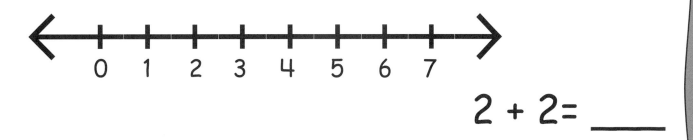

2 + 2= _____

Numeral Guide

1 2 3 4 5 6 7 8 9

Use the number line to show 3 + 1, then write the sum.

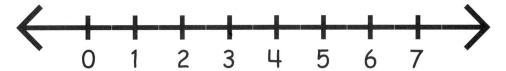

3 + 1= ___

Use the number line to show 3 + 2, then write the sum.

3 + 2= ___

Use the number line1to show 2 + 3, then write the sum.

2 + 3= ___

Use the number line to show 4 + 1, then write the sum.

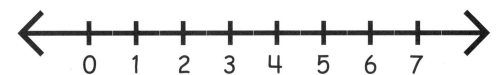

4 + 1= ___

Use the number line to show 1 + 4, then write the sum.

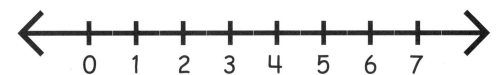

1 + 4= ___

Which plate has <u>fewer</u> cookies?

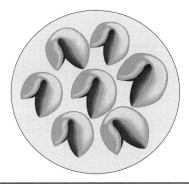 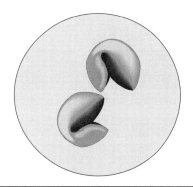

Which nest has <u>fewer</u> eggs?

Which vase has <u>fewer</u> flowers?

Which bowl has <u>fewer</u> fish?

Use the number line to show 1 + 3, then write the sum.

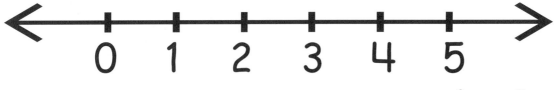

1 + 3= ___

Use the number line to show 1 + 4, then write the sum.

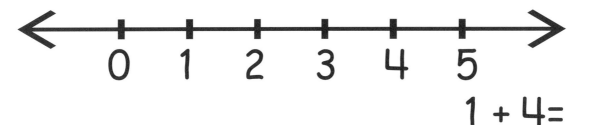

1 + 4= ___

Use the number line to show 2 + 1, then write the sum.

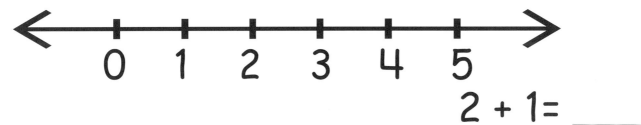

2 + 1= ___

Use the number line to show 4 + 1, then write the sum.

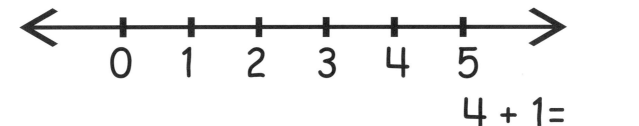

4 + 1= ___

Use the number line to show 3 + 2, then write the sum.

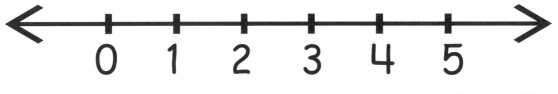

3 + 2= ___

Trace each numeral, then circle the sum of each group.

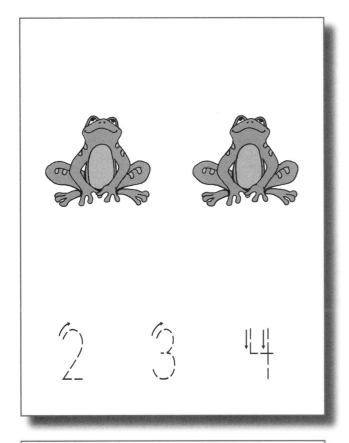

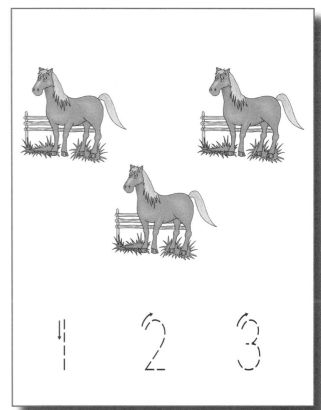

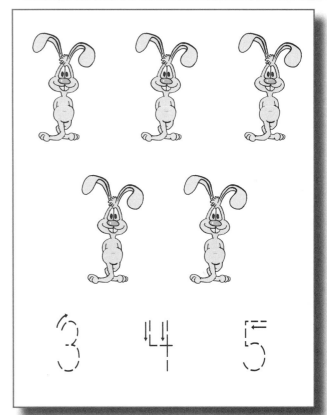

Draw, say, or point to the objects that would repeat the pattern.

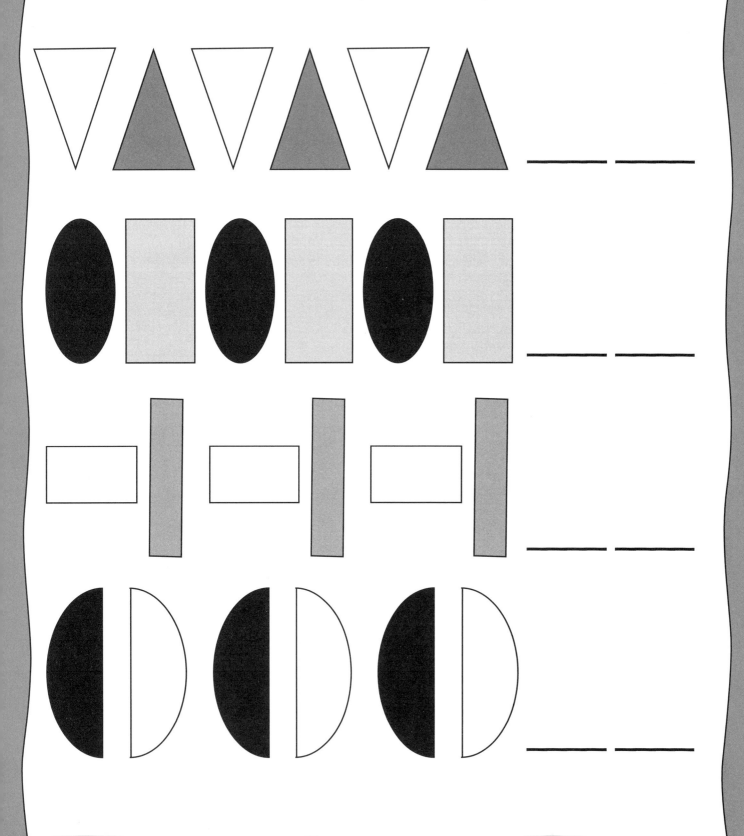

Draw, say, or point to the objects that would repeat the pattern.

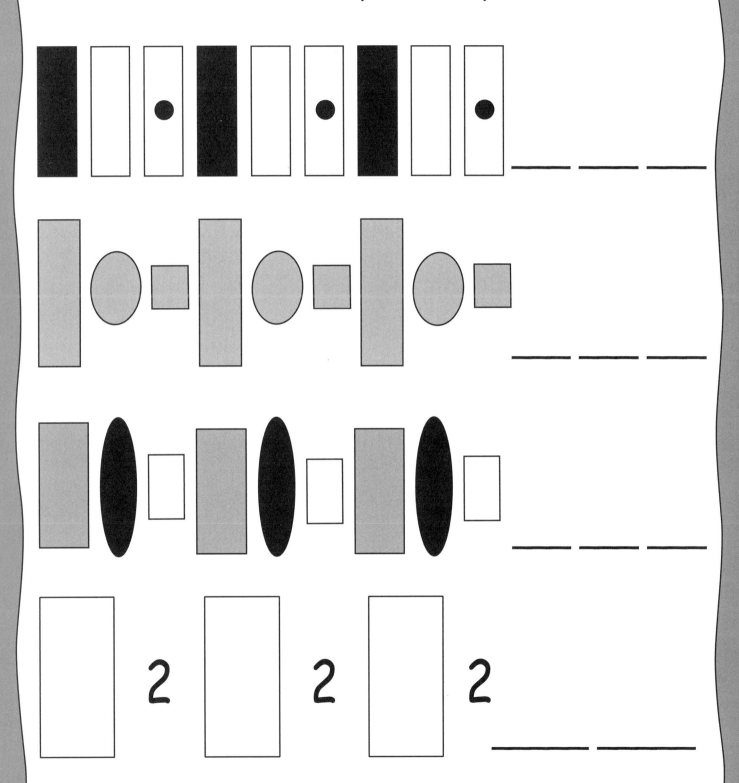

Use the clues to draw a line segment to match each child with their favorite animal at the zoo. Write the number of favorite animals in each group.

Tina named her favorite animals Ed, Eddie, and Edward.

_____

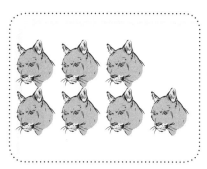

Beth has the most favorite animals.

_____

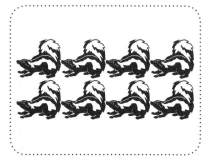

Ray has two more favorite animals than Tina.

_____

Tess has one less favorite animal than Beth.

_____

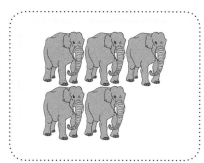

Numeral Guide

1 2 3 4 5 6 7 8 9

The Kitchen Puzzle

1. How many people are going to be eating at the table today? How do you know?

2. Will anyone eat fruit today? How do you know?

3. What will the people drink? How do you know?

4. Does the puzzle take place in the morning? How do you know?

5. Is there an animal that lives in this house? How do you know?

Write the missing numerals.

$1 + 2 = \underline{}$

$2 + \underline{} = 3$

$\underline{} + 2 = 5$

Write or say the total number of circles.

○○ + ○ = ___

○○ + ○○ = ___

○ + ○○ = ___

○○○ + ○ = ___

○○ + ○ = ___

Write or say the answers.

How many apples are there on both trees?

How many apples are there on both trees?

How many apples are there on both trees?

How many apples are there on both trees?

Write or say the sum of the oranges on the trees.

Count the tally marks in each set and say or write the total number of tallies in the space.

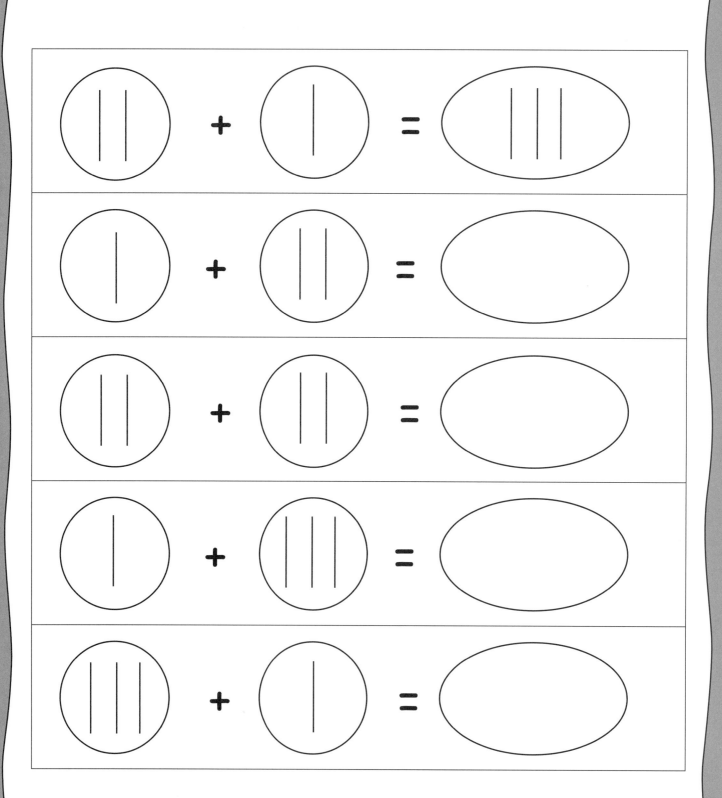

Count the tally marks in each set and say or write the total number of tallies in the space.

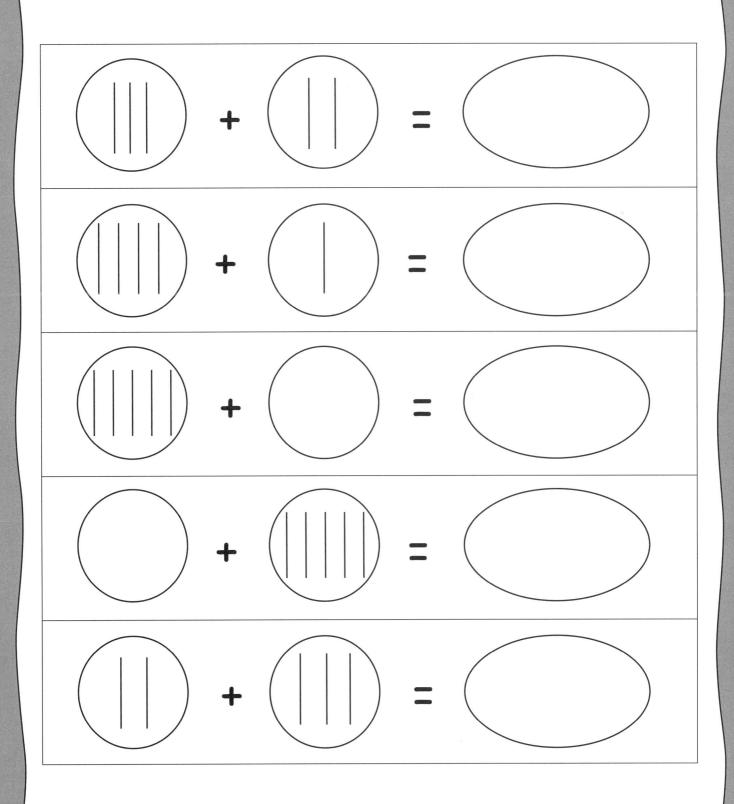

Can You Find Me?™

I am hiding next to shapes.
Can you figure out where?
I am under a circle
and between two squares.

Of the four pictures that you see,
tell me now, can you find me?

*For more activities like this, please see our Can You Find Me?™ Clues and Chooses series.

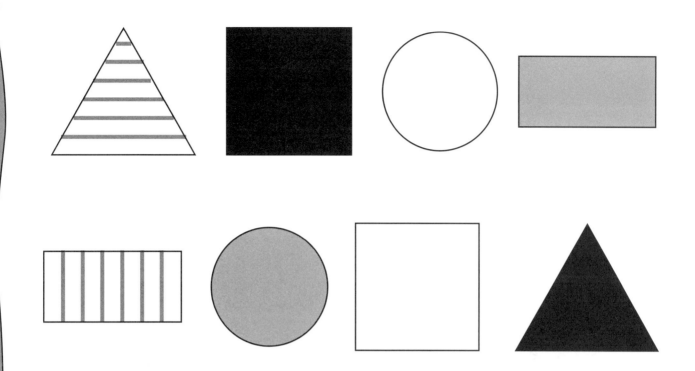

1. Point to the object that is black and not square.

2. Point to the object that is round and not white.

3. Point to the object that is gray without corners.

4. Point to the object that is striped with 4 corners.

5. Point to the object that has less than 3 corners and is white.

6. How many objects have no corners? _____

7. How many objects have less than 4 corners? _____

Write or say the sum of the apples on the trees.

Count the tally marks in each set and say or write the total number of tallies in the space.

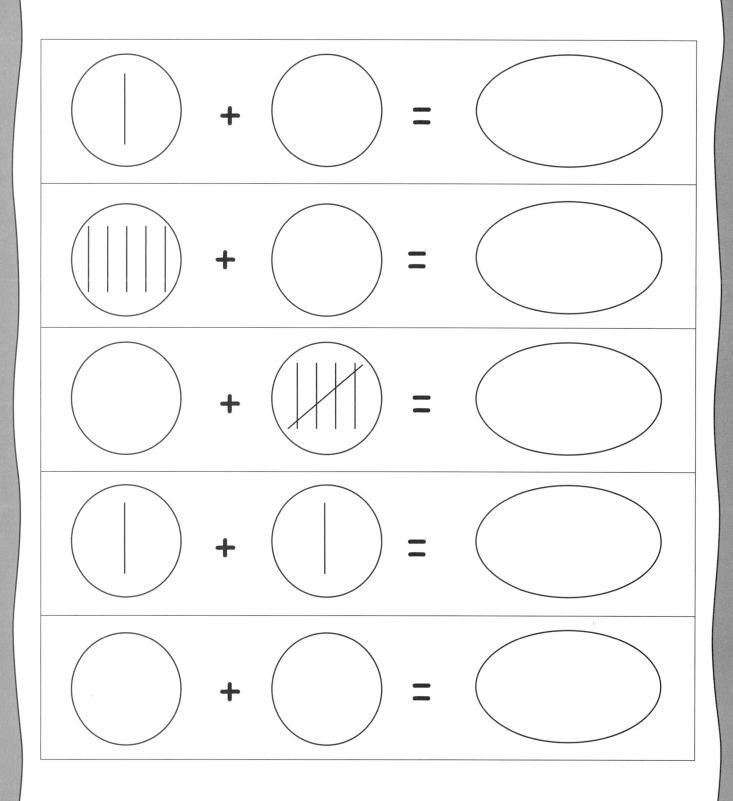

Count the tally marks in each set and say or write the total number of tallies in the space.

Write the total number.

☐ + 2 = _____

2 + ☐ = _____

☐☐ + ☐☐ = _____

☐☐☐ + ☐ = _____

☐ + ☐ = _____

Write the total number.

$\blacksquare\blacksquare$ + 1 = _____

$\blacksquare\blacksquare$ + 0 = _____

$\blacksquare\blacksquare\blacksquare$ + 0 = _____

$\blacksquare\blacksquare\blacksquare$ + 0 = _____

$\blacksquare\blacksquare$ + 0 = _____

If you added another apple to the bowl, how many apples would there be?

_____ +1 = _____

_____ +1 = _____

_____ +1 = _____

_____ +1 = _____

Write each sum, then say each number sentence.

$$0 \quad + \quad 1 \quad = \quad \underline{1}$$

Zero plus one equals one.

$$1 \quad + \quad 0 \quad = \quad \underline{}$$

$$1 \quad + \quad 1 \quad = \quad \underline{}$$

$$0 \quad + \quad 2 \quad = \quad \underline{}$$

$$1 \quad + \quad 2 \quad = \quad \underline{}$$

$$0 \quad + \quad 3 \quad = \quad \underline{}$$

$$1 \quad + \quad 3 \quad = \quad \underline{}$$

$$2 \quad + \quad 2 \quad = \quad \underline{}$$

$$0 \quad + \quad 4 \quad = \quad \underline{}$$

$$1 \quad + \quad 4 \quad = \quad \underline{}$$

$$1 \quad + \quad 5 \quad = \quad \underline{}$$

DEER

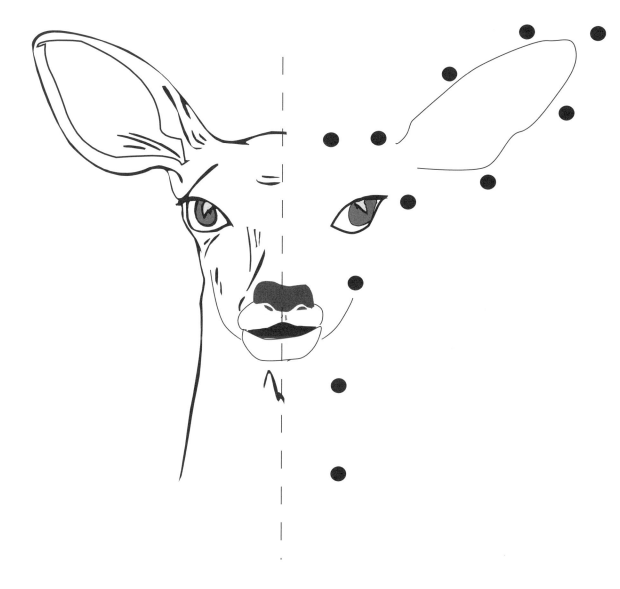

I am a fast runner. I can jump high over fences and bushes.

Draw the missing parts of the picture, then color the picture. Can you draw some grass for me to eat? Can you add something else to the picture?

*For more activities like this, please see our *Thinker Doodles™* Hafl and Half Animal series.

THINKER DOODLES™

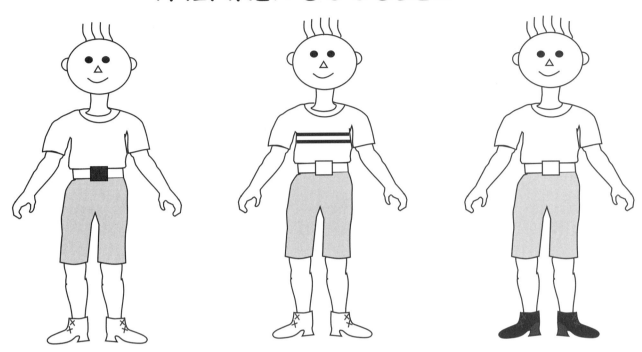

1. Look at each boy above, then find his unfinished picture below. Use a pencil to fill in all the missing parts.

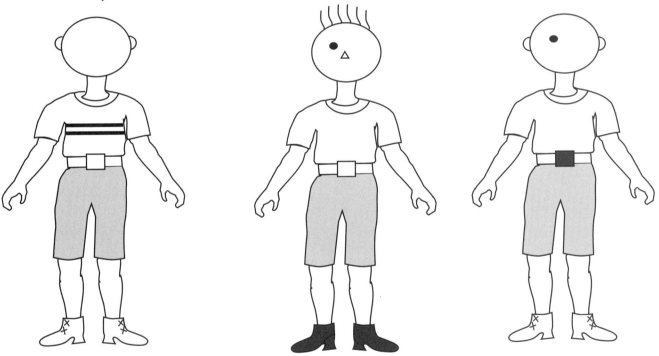

2. Color both boys with black heels with 2 colors.

3. Color both boys with black belt buckles with 3 colors.

4. Color both of the remaining boys with 1 color.

*For more activities like this, please see our *Thinker Doodles™* Clues and Chooses series.

Write the answer to each problem in the space provided.

1 + 2 = ____	2 + 2 = ____
3 + 2 = ____	0 + 4 = ____
0 + 3 = ____	3 + 1 = ____
1 + 3 = ____	2 + 2 = ____
5 + 0 = ____	0 + 5 = ____
1 + 1 = ____	3 + 0 = ____
4 + 1 = ____	2 + 3 = ____
2 + 1 = ____	1 + 0 = ____
1 + 4 = ____	2 + 2 = ____
3 + 2 = ____	2 + 0 = ____

Complete the number sentence
for each problem.

You have 2 toys and get 1 more. How many toys do you have after that?

_____ + _____ = _____

There are 3 books on one shelf and 2 more books on another shelf. How many books are there in all?

_____ + _____ = _____

You ate 1 cherry. Then you ate 4 more cherries. How many cherries did you eat?

_____ + _____ = _____

You have your bike. A friend brings another bike to your house. How many bikes are at your house?

_____ + _____ = _____

You have 2 pennies. You find 1 penny. Later you find another penny. How many pennies do you have?

_____ + _____ + _____ = _____

Point to the object or objects that answer(s) each question.

1. What object is round and not white?

2. What objects are black and not round?

3. What objects are white with corners?

4. What object is gray and has two other colors?

5. What objects are square and not striped?

1. I am a number less than eight but greater than six. Which number am I? _____

2. I am a number greater than 3 but less than 5. What number am I?

   _____

3. What two numbers are less than 7 but greater than 4? _____

4. What numbers are greater than zero but less than 4? _____

5. Circle all the numbers less than 9 but greater than 5.

   ## 4 5 6 7 8 9 10

6. Circle all the numbers greater than 2 but less than 8.

   ## 1 2 3 4 5 6 7 8 9 10

Write the answer to each problem in the space provided.

$1 + 1 =$ _____	$3 + 0 =$ _____
$1 + 2 =$ _____	$2 + 2 =$ _____
$4 + 1 =$ _____	$2 + 3 =$ _____
$0 + 3 =$ _____	$3 + 1 =$ _____
$1 + 3 =$ _____	$2 + 2 =$ _____
$3 + 2 =$ _____	$0 + 4 =$ _____
$2 + 1 =$ _____	$1 + 0 =$ _____
$1 + 4 =$ _____	$2 + 2 =$ _____
$5 + 0 =$ _____	$0 + 5 =$ _____
$3 + 2 =$ _____	$2 + 0 =$ _____

Complete the number sentence for each problem.

A pet store has 2 dogs and gets 1 more. How many dogs does the pet store have after that?

_____ + _____ = _____

There is 1 book on one shelf and 1 more book on another shelf. How many books are there in all?

_____ + _____ = _____

You ate 2 cherries. Then you ate 2 more cherries. How many cherries did you eat?

_____ + _____ = _____

You have your 3 marbles. A friend brings you 2 more marbles. How many do you have now?

_____ + _____ = _____

You have 1 penny. You find 2 pennies. Later you find another penny. How many pennies do you have?

_____ + _____ + _____ = _____

Beginning Subtraction

How many oranges
are on the tree? _____

You eat one orange.
Put an X on the one you ate.

How many oranges are left? _____

How many apples
are on the tree? _____

You eat two apples.
Put an X on the one you ate.

How many apples are left? _____

How many oranges
are on the tree? _____

You eat three oranges.

How many oranges are left? _____

How many apples
are on the tree? _____

You eat three apples.

How many apples are left? _____

How many oranges are on the tree?

_____

You eat this many. __5__ (cross them out)

How many oranges are left? _____

You eat this many apples off the tree. __1__ (cross them out)

How many apples are left?

_____

You eat this many oranges off the tree. __5__ (cross them out)

How many oranges are left?

_____

You eat this many apples off the tree. __6__ (cross them out)

How many apples are left?

_____

Draw, write, and say the answer for each of the following subtraction problems.

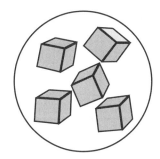 take away 1 = _____

(cross it out)

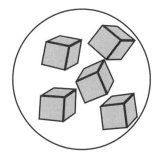 take away 5 = _____

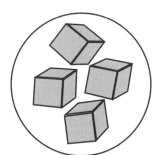 take away 3 = _____

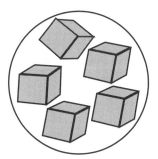 take away 1 = _____

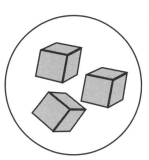 take away 1 = _____

Note: It sometimes helps students if they cross out the items taken away from the original amount.

Draw, write, and say the answer for each of the following subtraction problems.

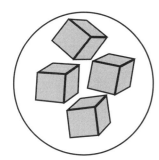　　take away 3 = _____

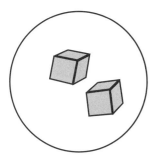　　take away 2 = _____

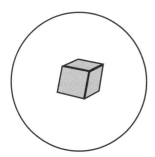　　take away 1 = _____

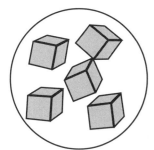　　take away 2 = _____

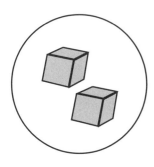　　take away 1 = _____

How many peas are on the plate?

You eat this many. Cross this many out.

How many peas are left on the plate?

_____ - __3__ = _____

How many apple slices are on the plate?

You eat this many.

How many are left on the plate?

_____ - __4__ = _____

How many carrots are on the plate?	You eat this many. Cross them out.	Write how many are left.
↓	↓	↓

_____ − __3__ = _____

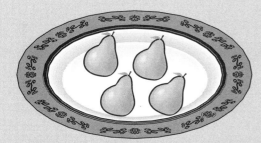

How many pears are on the plate?	You eat this many.	Write how many are left.
↓	↓	↓

_____ − __4__ = _____

How many plums are on the plate?	You eat this many.	Write how many are left.
↓	↓	↓

_____ − __2__ = _____

How many bananas? | You eat this many. | How many are left?

_____ - 1 = _____

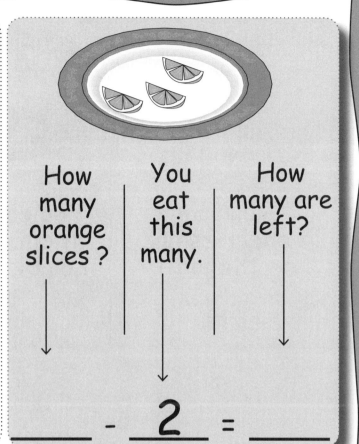

How many orange slices ? | You eat this many. | How many are left?

_____ - 2 = _____

How many eggs? | You eat this many. | How many are left?

_____ - 1 = _____

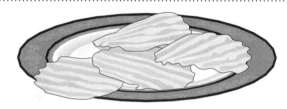

How many potato chips? | You eat this many. | How many are left?

_____ - 1 = _____

How many cherries are on the plate?

You eat two of them. Put the number you ate here.

How many cherries are left?

___ − ___ = ___

How many tomatoes are on the plate?

You eat one of them. Put the number you ate here.

How many are left?

___ − ___ = ___

1. The black house is farthest from which house?

2. Which truck is closest to the white house?

3. Which truck is farthest from the gray house?

4. What is the third house that the black truck will pass on the road?

5. What is the second house that the gray truck will pass on the road?

MIND BENDERS®

DIRECTIONS: Fill in the chart using Y for yes or N for no as you solve the puzzle.

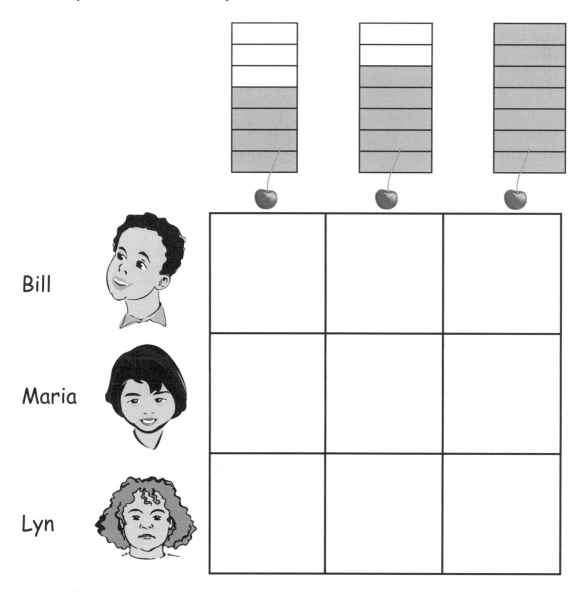

Bill

Maria

Lyn

Bill, Maria, and Lyn all ate cherries for dessert. Use the clue and chart to find how many cherries each person ate.

1. Bill had more than Lyn but less than Maria.

Complete each number sentence below by crossing out the subtracted items on the plate before writing the solution.

$$4 - 3 = \rule{1cm}{0.4pt}$$

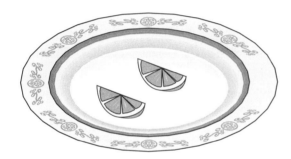

$$2 - 1 = \rule{1cm}{0.4pt}$$

$$5 - 2 = \rule{1cm}{0.4pt}$$

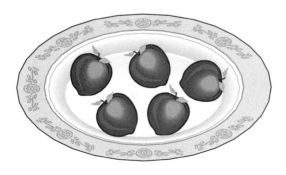

$$5 - 5 = \rule{1cm}{0.4pt}$$

Write and say the answer to each of the following subtraction problems.

 $4 - 2 = \underline{}$

 $5 - 2 = \underline{}$

 $5 - 4 = \underline{}$

 $1 - 1 = \underline{}$

A number line can be used to show subtraction. Starting at zero, hop to the first number. Starting there, hop back the number being subtracted to find the answer. Use the number line to show the solution to each number sentence.

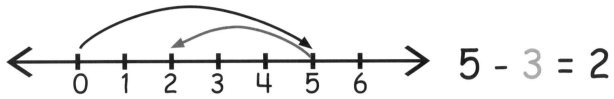

$5 - 3 = 2$

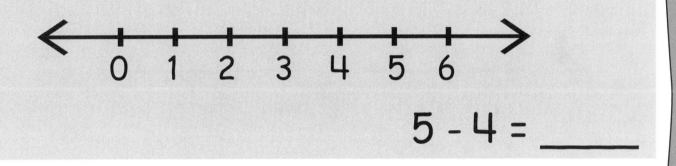

$5 - 4 = \underline{\qquad}$

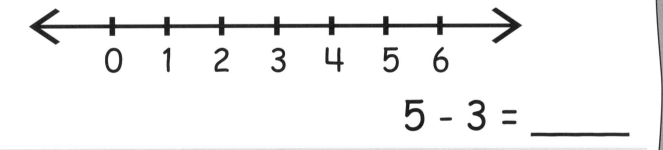

$5 - 3 = \underline{\qquad}$

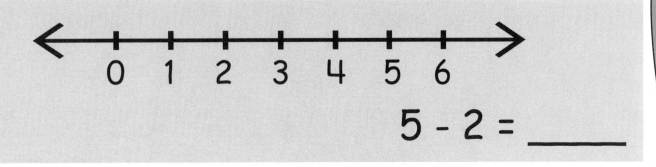

$5 - 2 = \underline{\qquad}$

Use the number line to show the solution to each number sentence.

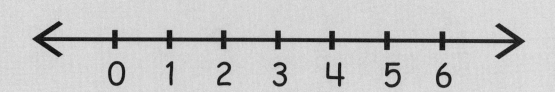

4 - 2 = _____

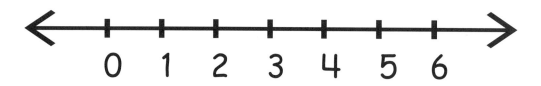

4 - 3 = _____

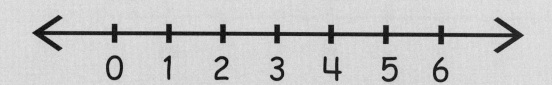

3 - 2 = _____

2 - 1 = _____

Use the number line to show the solution to each number sentence.

4 - 1 = _____

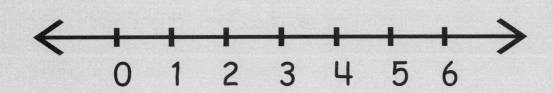

5 - 1 = _____

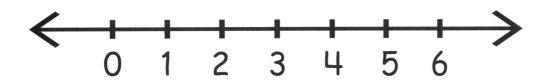

3 - 1 = _____

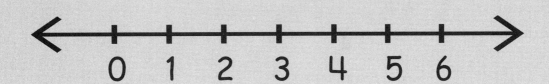

2 - 1 = _____

THINKER DOODLES™

1. Look at each space shuttle above, then find its unfinished picture below. Use a pencil to draw in all the missing parts.

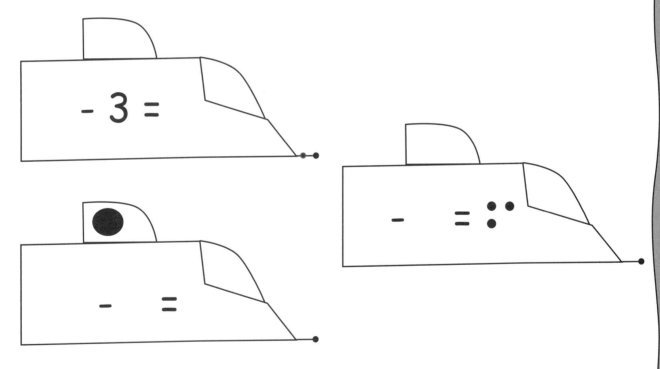

2. Color both shuttles with a difference of 4 black dots with two colors.

3. Color both shuttles with the fewest black dots with three colors.

4. Color both of the remaining shuttles different from all the others.

*For more activities like this, please see our *Thinker Doodles™* Clues and Chooses series.

Draw a line segment to each set of objects that made each picture.

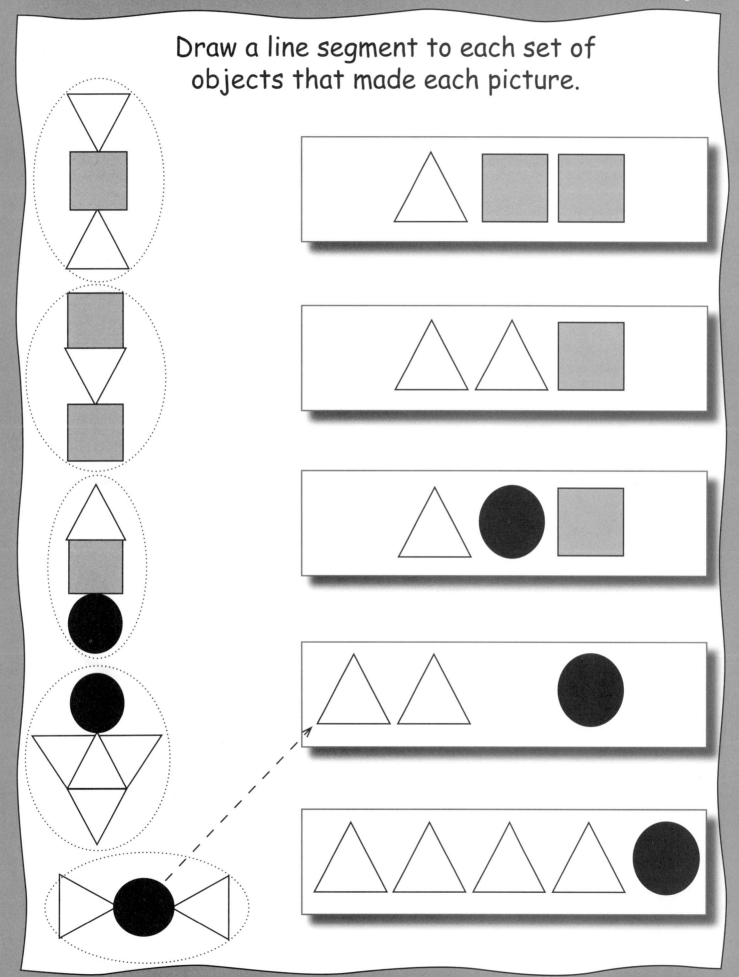

Use the number line to show the solution to each number sentence.

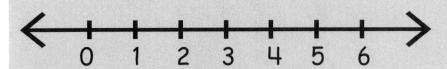

$3 - 2 = $ _____

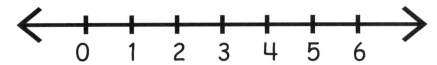

$4 - 3 = $ _____

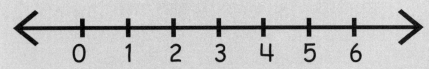

$5 - 4 = $ _____

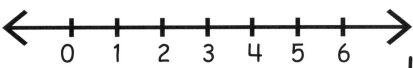

$4 - 2 = $ _____

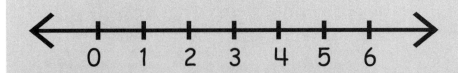

$4 - 1 = $ _____

$3 - 1 = $ _____

Use the number line to show the solution to each number sentence.

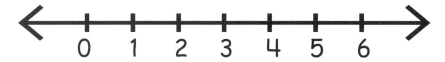

$2 - 1 =$ _____

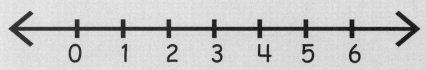

$4 - 3 =$ _____

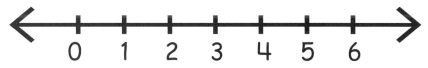

$5 - 0 =$ _____

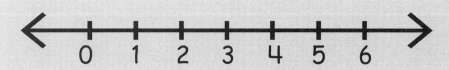

$5 - 2 =$ _____

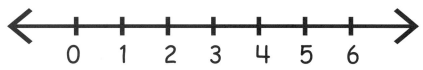

$4 - 2 =$ _____

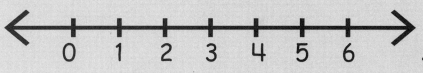

$3 - 2 =$ _____

Complete the tally marks to find the solution to each number sentence.

take away ... leaves

$$2 - 1 = \underline{\hspace{2cm}}$$

take away ... leaves

$$3 - 1 = \underline{\hspace{2cm}}$$

take away ... leaves

$$5 - 1 = \underline{\hspace{2cm}}$$

take away ... leaves

$$4 - 2 = \underline{\hspace{2cm}}$$

take away ... leaves

$$4 - 4 = \underline{\hspace{2cm}}$$

Complete the tally marks to find the solution to each number sentence.

 take away leaves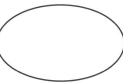

$$3 - 2 = \underline{\hspace{2cm}}$$

 take away leaves

$$5 - 3 = \underline{\hspace{2cm}}$$

 take away leaves

$$5 - 5 = \underline{\hspace{2cm}}$$

 take away leaves

$$4 - 3 = \underline{\hspace{2cm}}$$

 take away leaves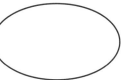

$$1 - 0 = \underline{\hspace{2cm}}$$

Write the answer to each number sentence.

4 - 2 = _____	5 - 2 = _____
3 - 2 = _____	4 - 1 = _____
5 - 3 = _____	5 - 1 = _____
5 - 0 = _____	4 - 2 = _____
3 - 1 = _____	5 - 5 = _____
4 - 1 = _____	3 - 2 = _____
2 - 1 = _____	4 - 3 = _____
5 - 4 = _____	1 - 0 = _____
3 - 2 = _____	2 - 2 = _____
4 - 3 = _____	2 - 0 = _____

Venn Diagrams

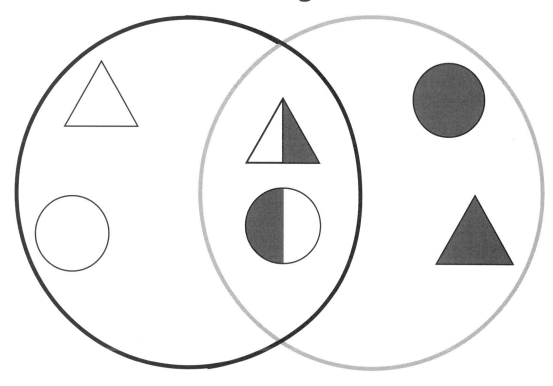

1. Describe the color of the objects that are in both the gray and black circle.

2. What is the same about all the objects in the black circle?

3. What is the same about all the objects in the gray circle?

4. Point to where each object below belongs in the picture.

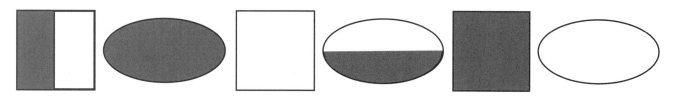

1. Circle the shape that is most like the pie.

2. Circle the shape that is most like the sandwich.

3. Circle the shape that is most like the dollar bill.

4. Circle the shape that is most like the dolphin's top fin.

Write the answer to each number sentence.

3 - 2 = _____	5 - 3 = _____
2 - 2 = _____	3 - 1 = _____
5 - 5 = _____	1 - 1 = _____
4 - 1 = _____	4 - 3 = _____
5 - 1 = _____	2 - 2 = _____
5 - 4 = _____	4 - 0 = _____
4 - 3 = _____	2 - 0 = _____
3 - 1 = _____	3 - 2 = _____
3 - 0 = _____	4 - 4 = _____
4 - 1 = _____	1 - 0 = _____

Write each number sentence.

You have 5 toys. You lose 3. How many toys do you have after that?

_____ - _____ = _____

You have 4 books on the shelf in your room. You return 2 to the library. How many books are left on the shelf?

_____ - _____ = _____

There are 5 cherries. You ate 2 cherries. How many cherries were not eaten?

_____ - _____ = _____

You have 5 pennies. You spend 5 pennies. How many pennies do you have left?

_____ - _____ = _____

You took 5 swings at balls on a batting T. You get 4 hits. How many times did you miss the ball on the batting T?

_____ - _____ = _____

Write and say the answer to each of the following subtraction problems.

 4 - 1 = __

 5 - 3 = __

 5 - 2 = __

 1 - 0 = __

A store has 2 dogs and 2 cats. A bar graph that
shows this could look like this:

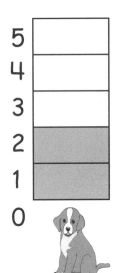

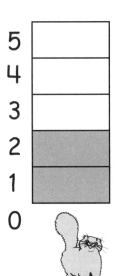

A farmer has 4 pigs and 1
horse. Create a bar graph
that shows this.

A farmer has 3 goats and
5 sheep. Create a bar graph
that shows this.

Write and say each
number sentence.

You have 4 toys. You lose 3. How many toys do you have after that?

_____ - _____ = _____

You have 3 books on the shelf in your room. You return 2 to the library. How many books are left on the shelf?

_____ - _____ = _____

There are 5 cherries. You ate 4 cherries. How many cherries were not eaten?

_____ - _____ = _____

You have 3 pennies. You spend 3 pennies. How many pennies do you have left?

_____ - _____ = _____

You took 5 swings at balls on a batting T. You get 0 hits. How many times did you miss the ball on the batting T?

_____ - _____ = _____

A store has 3 snakes and 2 bats. Create a bar graph that shows this.

5
4
3
2
1
0

5
4
3
2
1
0

A child has 3 white socks and 4 black socks. Create a bar graph that shows this.

5
4
3
2
1
0

5
4
3
2
1
0

A child has 1 teddy bear and 3 dolls. Create a bar graph that shows this.

5
4
3
2
1
0

5
4
3
2
1
0

A child has 3 crayons and 2 balls. Create a bar graph that shows this.

5
4
3
2
1
0

5
4
3
2
1
0

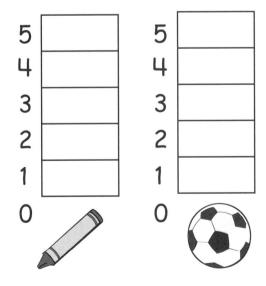

How many of each animal is in the pet store?

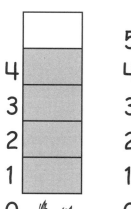 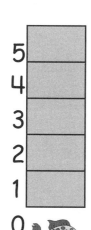 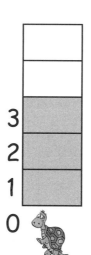

 _____

_____

How many of each car is there?

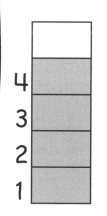  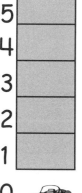 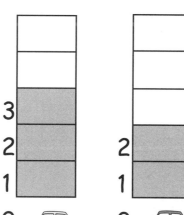

 _____

 _____

_____

How many of each candy was eaten?

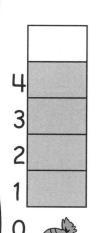 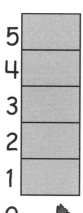 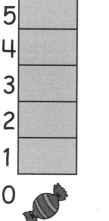 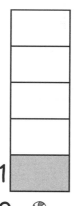 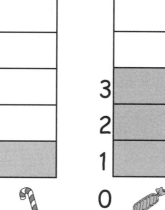

 _____

_____

Trace each numeral, then circle the correct answer.

How many children are in the picture?

5 6 7

How many dogs are in the picture?

3 4 5

How many cats are in the picture?

1 2 3

Point or say where the girl will probably go.

Point or say what the boy will probably do.

  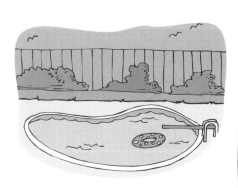

Point or say where the children will probably go.

A store has 4 owls and 2 raccoons.
Create a bar graph that shows this.

A ball player has 1 glove, 2 balls, and 1 bat. Create a bar graph that shows this.

A family has 1 car, 1 mini-van, 5 bikes, and 2 skateboards.
Create a bar graph that shows this.

A store has 3 dogs and 2 cats. Create a bar graph that shows this.

A family has 2 cars, 5 bicycles, and 3 skateboards. Create a bar graph that shows this.

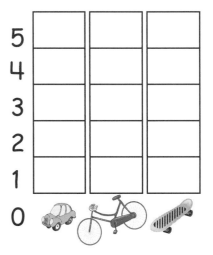

A child has 1 bicycle, 1 skateboard, 1 pair of skates, and 3 helmets. Create a bar graph that shows this.

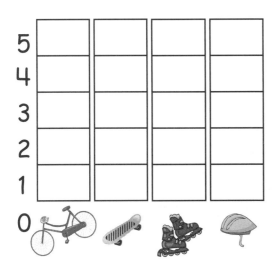

A toy box holds 3 balls, 1 bat, 2 gloves, and 4 helmets. Create a bar graph that shows this.

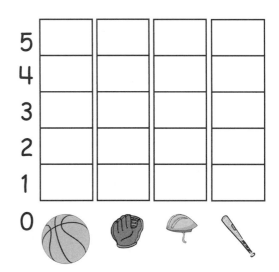

You have 4 white balls and 1 black ball. If you put them in a box, closed your eyes, and picked up one ball, what color would it probably be?

You have 5 white balls. If you put them in a box, closed your eyes, and picked up one ball, what color would it be?

You have 3 black balls and 1 gray ball. If you put them in a box, closed your eyes, and picked up one ball, what color would it probably be?

You have 2 gray balls and 2 black balls. If you put them in a box, closed your eyes, and picked up one ball, what color would it probably be?

You have 3 white socks and 1 black sock. If you put them in a box, closed your eyes, and picked up one sock, what color would it probably be, and why?

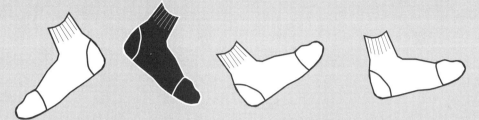

You have 5 gray books and 2 white books. If you put them in a box, closed your eyes, and picked up one book, what color would it probably be, and why?

You have 3 white socks, 2 gray socks, and 1 black sock. If you put them in a box, closed your eyes, and picked up one sock, what color would it probably be, and why?

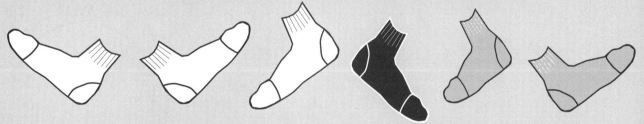

You have 1 white marble, 1 striped marble, 1 gray marble, and 3 black marbles. If you put them in a box, closed your eyes, and picked up one marble, what color would it probably be, and why?

Here is a whole cookie.

1

This cookie has been cut into
<u>two equal parts</u>, called halves ($\frac{1}{2}$).

2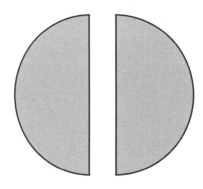

An object that is cut into two equal
parts is cut into halves.

Color $\frac{1}{2}$ of each object.

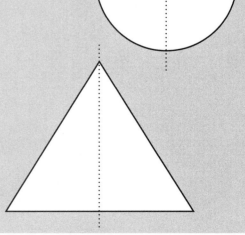

An object that is split into two parts of the same size has two halves. Circle each object that has one half of its area colored. Put an X through each object that does not.

Draw a line of symmetry to split each item in half, then color one of the halves.

Color most of the balls green and some of the balls red.

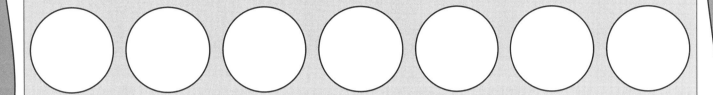

Color some of the balls blue, but not all the balls blue.

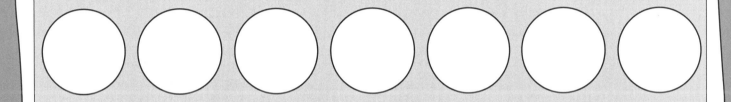

Leave most of the balls white, but not all the balls white.

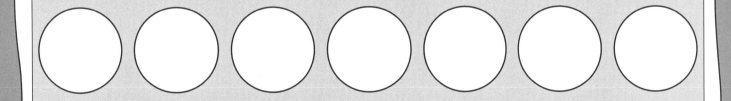

Color most of the balls red, but some of the balls blue.

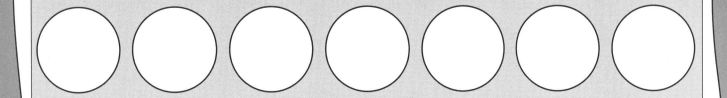

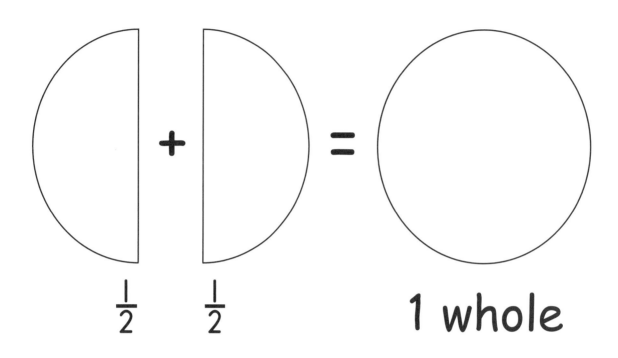

$$\frac{1}{2} \qquad \frac{1}{2} \qquad \text{1 whole}$$

Draw the missing part of the number sentence.

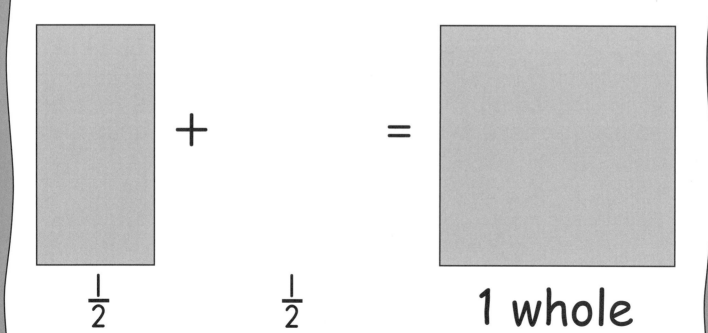

$$\frac{1}{2} \qquad \frac{1}{2} \qquad \text{1 whole}$$

Can You Find Me?™

Kyle ate half a pie,
then left half for Ron.
Ron ate half of what was left,
and gave the rest to Don.

Look at the pictures of the tasty treat,
then point to the piece that Don got to eat.

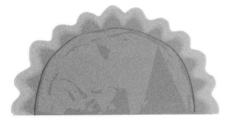

*For more activities like this, please see our *Can You Find Me?™* Clues and Chooses series.

Say, draw, or point to the object that would continue the pattern.

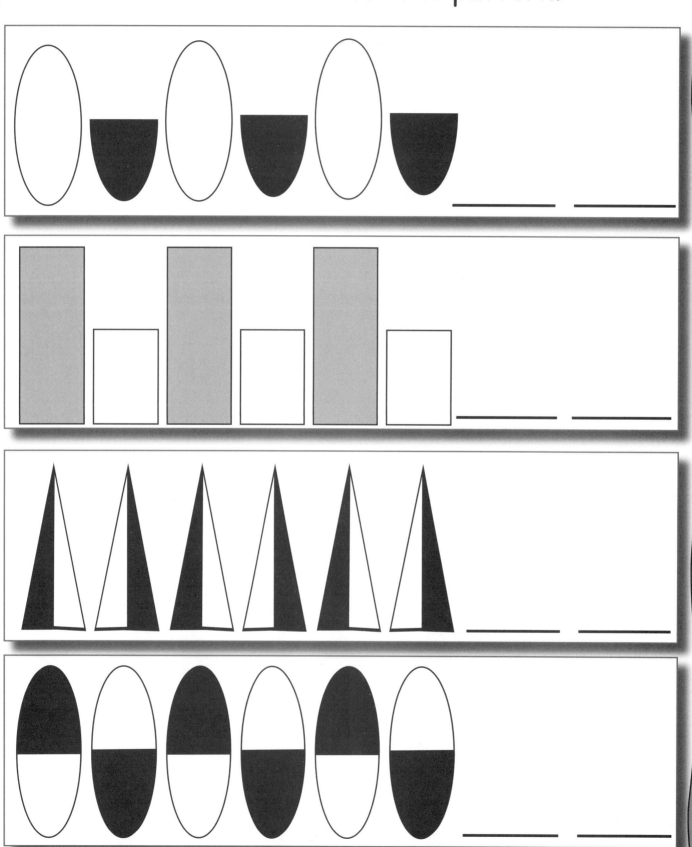

If you have 5 pennies, could you buy all of this candy?

If you have 4 pennies, could you buy all of this candy?

If you have 2 pennies, could you buy all of this candy?

If you have 3 pennies, could you buy all of this candy?

Our coins have names.

1¢   5¢   10¢   25¢

penny   nickel   dime   quarter

Say the name of each coin.

 1¢    5¢

 10¢    25¢

Count from 1 to 5.

What missing number is <u>between</u> 3 and 5?

3 __ 5

What number is <u>between</u> 2 and 4?

2 __ 4

What number is <u>between</u> 1 and 3?

1 __ 3

What number is <u>between</u> 4 and 6?

4 __ 6

Trace each numeral, then draw a line segment to the matching picture.

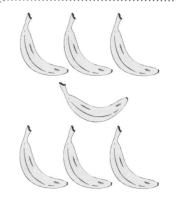

Trace each numeral, then draw a line segment to the matching picture.

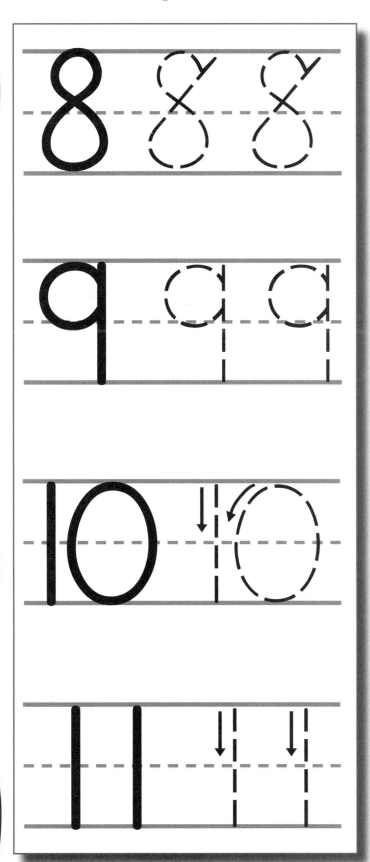

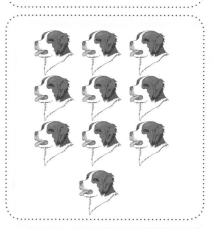

Trace each numeral, then draw a line segment to the matching picture.

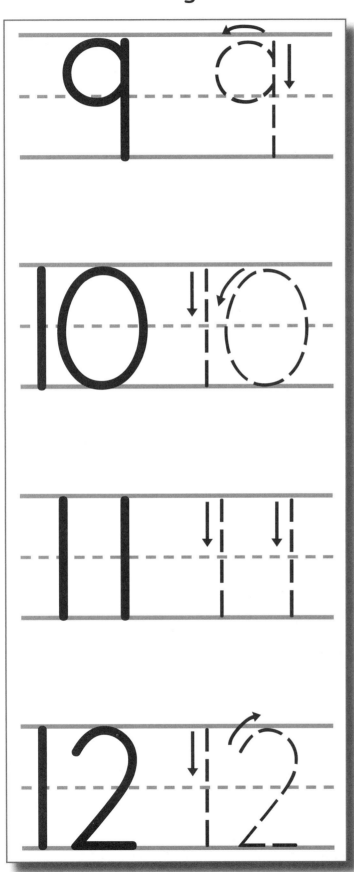

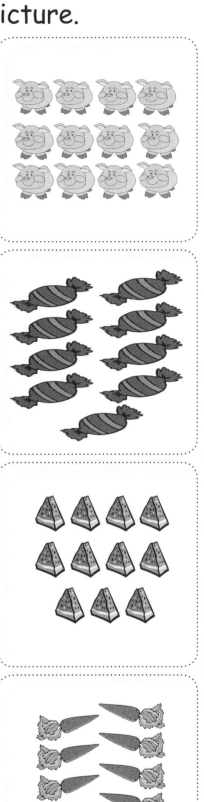

Touch the gray beads and say the numbers.

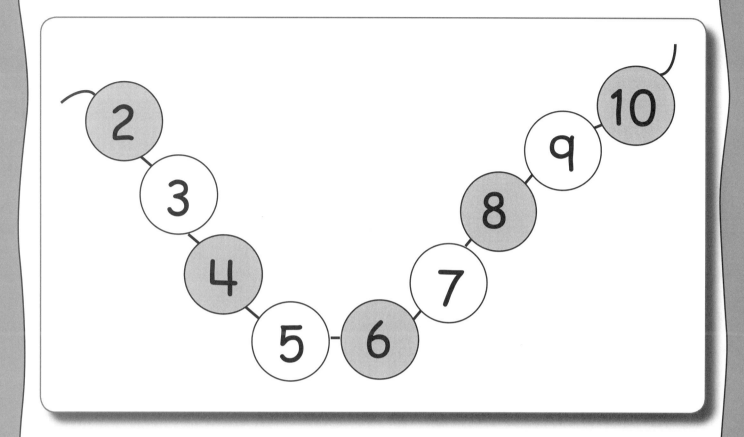

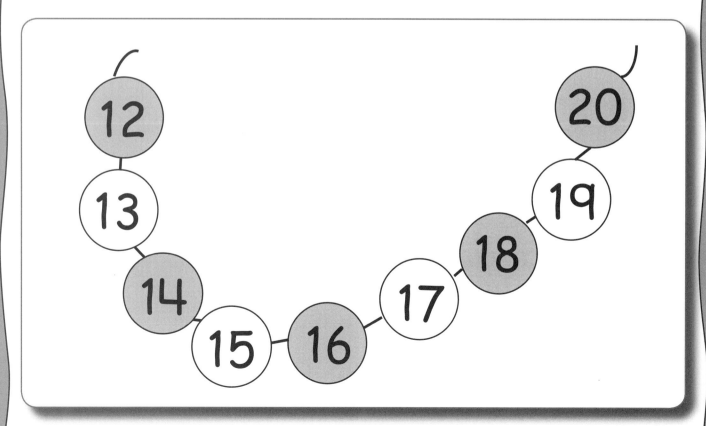

Explain whether the sun will probably shine or if it will probably rain.

Explain whether a boy is probably going to read or probably going to ride his bike.

Explain whether a girl will probably go to a party or probably play in a sand box.

Write or say the answer.

How many black faces? _____

Is the number even or odd?

_____

How many white faces? _____

Is the number even or odd?

_____

How many gray faces? _____

Is the number even or odd?

_____

How many black faces? _____

Is this number even or odd?

Which of the numbers are even? _____
Which of the numbers are odd? _____

Trace each numeral, then draw a line segment to the matching picture.

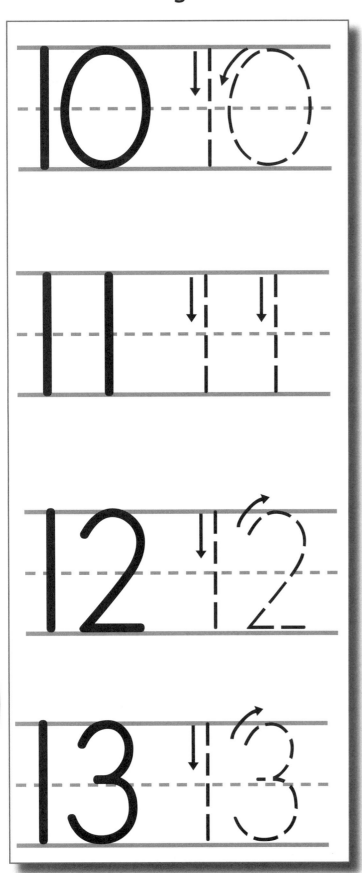

Trace each numeral, then draw a line segment to the matching picture.

Trace each numeral, then draw a line segment to the matching picture.

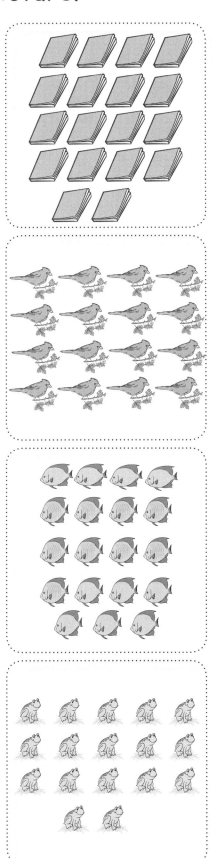

Draw, say, or point to the objects that would repeat the pattern.

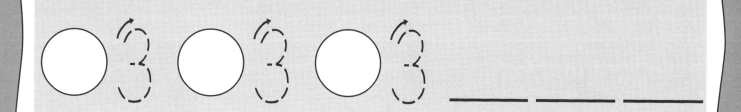

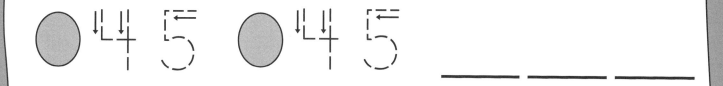

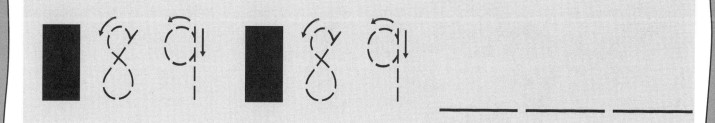

Draw, say, or point to the objects that would repeat the pattern.

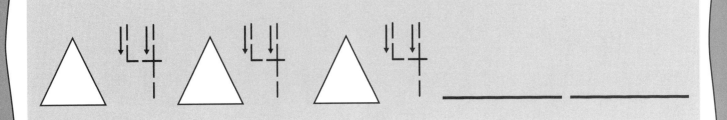

What time is it on this clock?
Is this the time you would probably eat breakfast?

What time is it on this clock?
Is this the time you would probably eat or play a
ball game on a weekday during the school year?

Which clock shows 1 o'clock?

Which clock shows 2 o'clock?

Which clock shows 4 o'clock?

Which clock shows 6 o'clock?

Which clock shows 8 o'clock?

Which clock shows 9 o'clock?

Draw the short hour hand of the clock to show the time.

one o'clock

five o'clock

eleven o'clock

nine o'clock

three o'clock

seven o'clock

What time is it on the clock?
Is this the time you would
probably go to bed?

What time is it on the clock?
Is this the time you would
probably go to bed?

What time is it on the clock?
Is this the time you would
probably go to bed?

What time is it on the clock?
Is this the time you would
probably go to bed?

Trace each numeral, then draw a line segment to its matching picture.

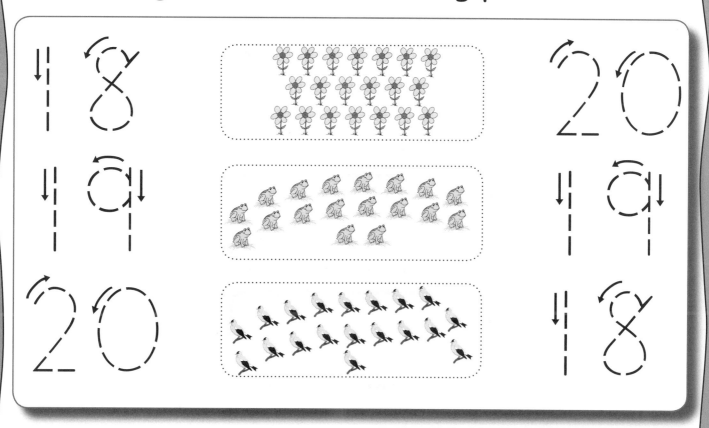

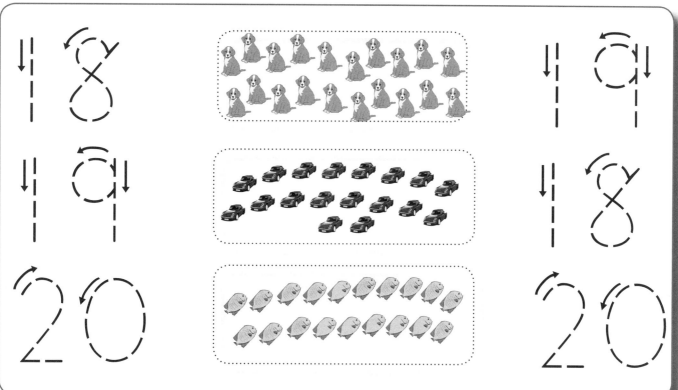

Trace each numeral, then draw a line segment to its matching picture.

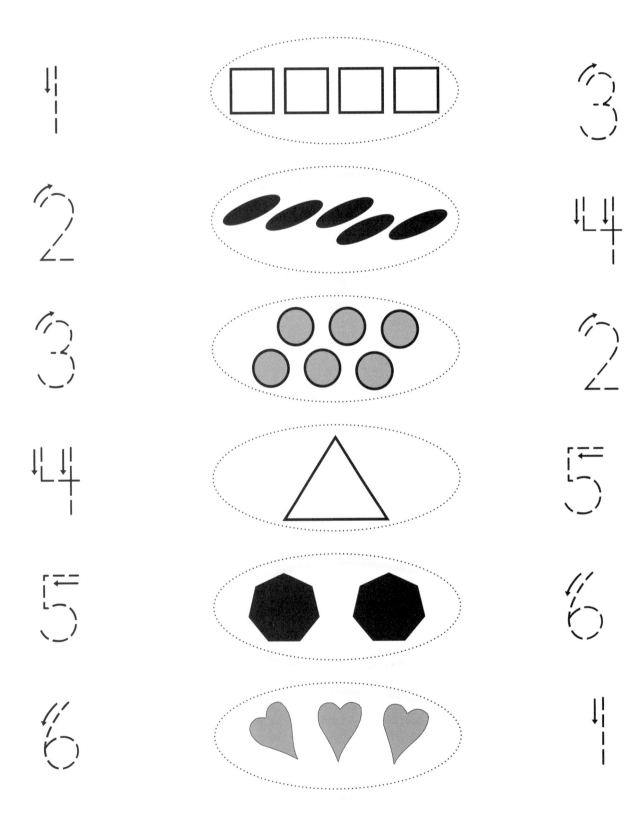

Touch and state the name of each shape, then follow the directions.

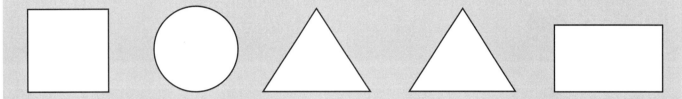

Draw a circle in all of the triangles.

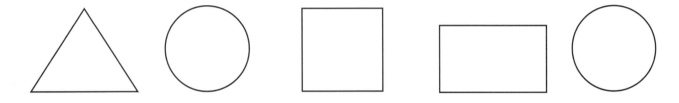

Draw a square in all of the circles.

Draw a triangle in all of the squares.

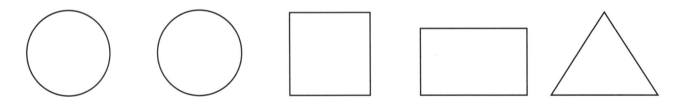

Draw a circle in all of the shapes that have corners.

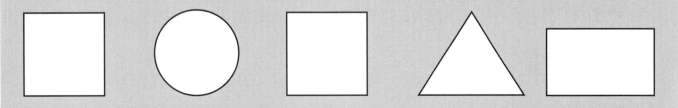

Draw a triangle in all of the rectangles.

NOTE- - Marking the square as a rectangle is acceptable since all squares are rectangles. Although, at this point, a child might not know all squares are rectangles.

Think About It

Which is longer, your arm or your foot?

Which is shorter, your bed or your leg?

Who is taller, your mom or you?

Which is shorter, a flagpole or the flag that goes on it?

Which is shorter, a baseball bat or your shoe?

Which is longer, your shoe or your arm?

Which is higher, the roof of a car or the seat on a bike?

Which takes longer, eating lunch or watching a movie?

Draw an X on all of the triangles.

Draw an X on all of the circles.

Draw an X on all of the squares.

Draw an X on all of the shapes that have corners.

Draw an X on all of the triangles.

Think About It

Which is longer, your arm or your height?

Which is shorter, your bed or your pant leg?

Who is shorter, your mom or you?

Which is taller, a toy car or a real car?

Which is shorter, a Christmas tree or a dandelion?

Which has the greatest distance across, a marble, a soccer ball, or a baseball?

Which is lower, the roof of a car, the roof of a bus, or the seat on a bike?

Which takes longer, eating lunch, sleeping at night, or reading a book?

Trace each numeral, then draw a line segment to its matching picture.

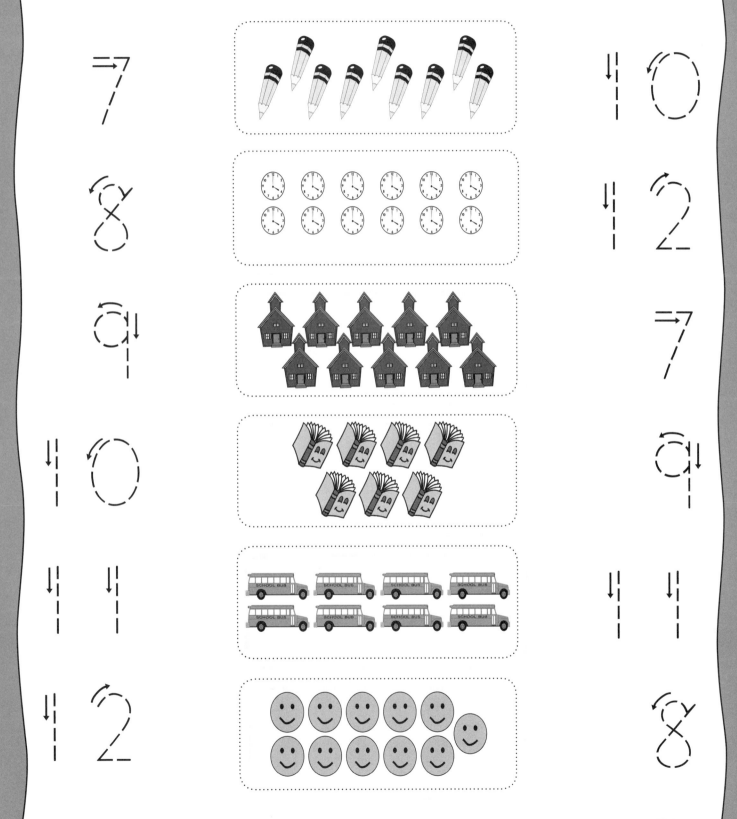

Touch and say the number on each white leaf.

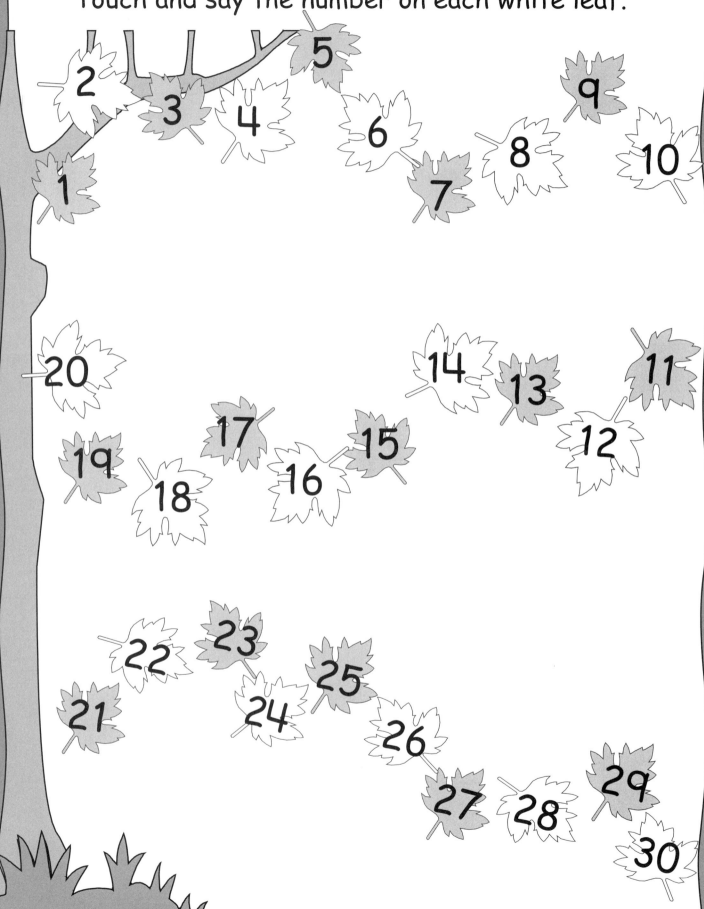

Eight bunnies were playing hide and seek with their mother. The mother found 1 bunny behind a flower box and 2 bunnies behind the bushes.

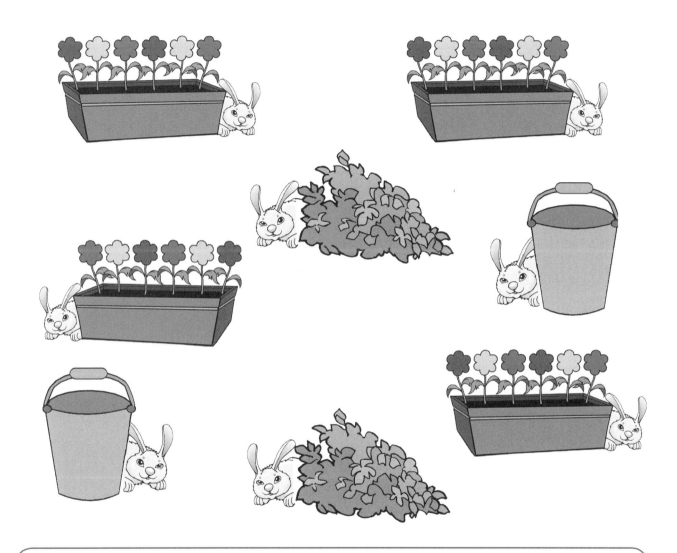

1. How many bunnies did the mother find? _____

2. How many bunnies were still hiding? _____

3. How many bunnies were still hiding behind the flower boxes? _____

Color $\frac{1}{2}$ of each shape.

Write each numeral, then say each number sentence.

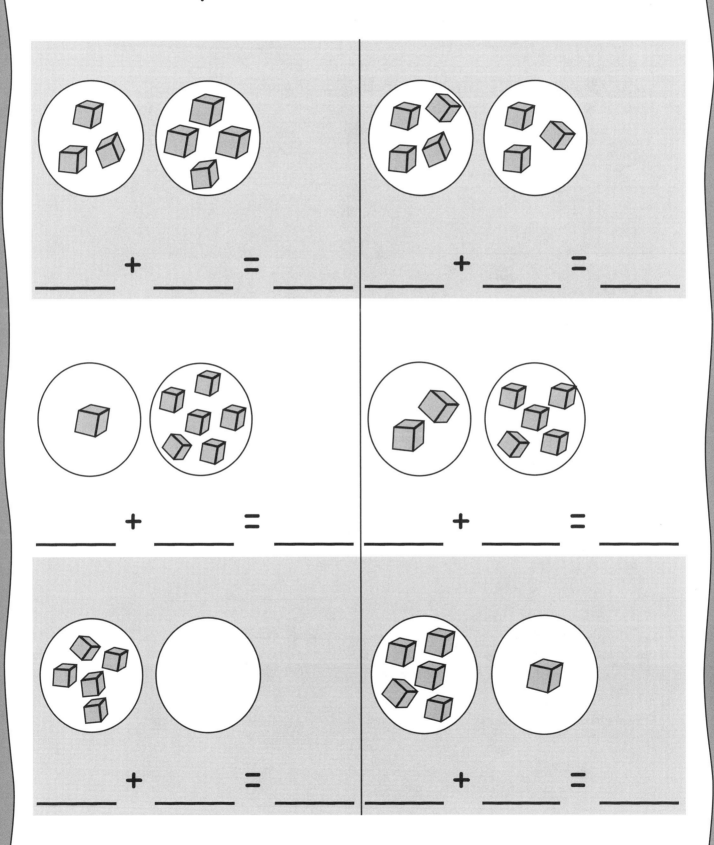

Write each numeral, then say each number sentence.

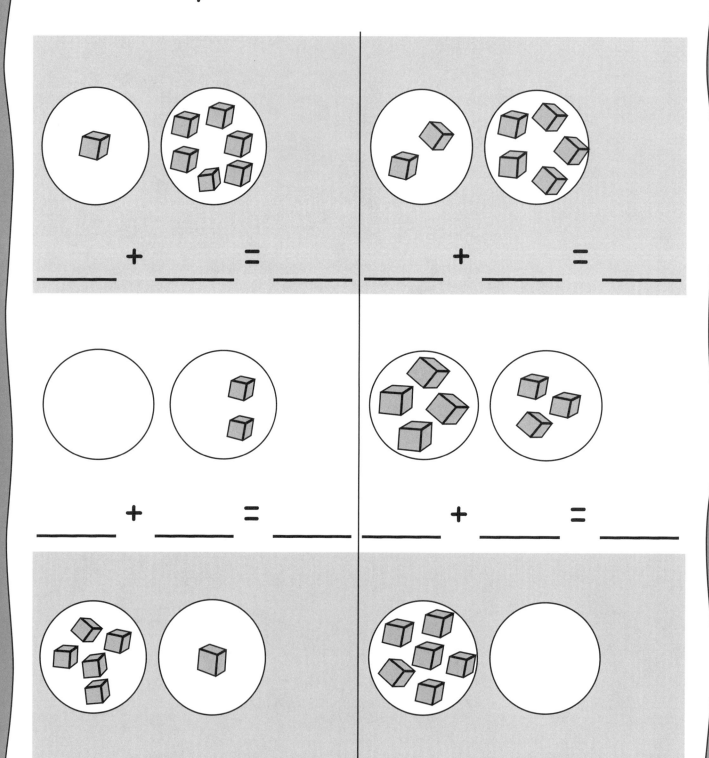

Begin with **1** . Touch each ball to count to ⑩.

Skip Counting: Use the white balls to count to ⑩ by twos (2, 4, 6, 8, 10). Begin at ②.

Write each numeral, then say each number sentence.

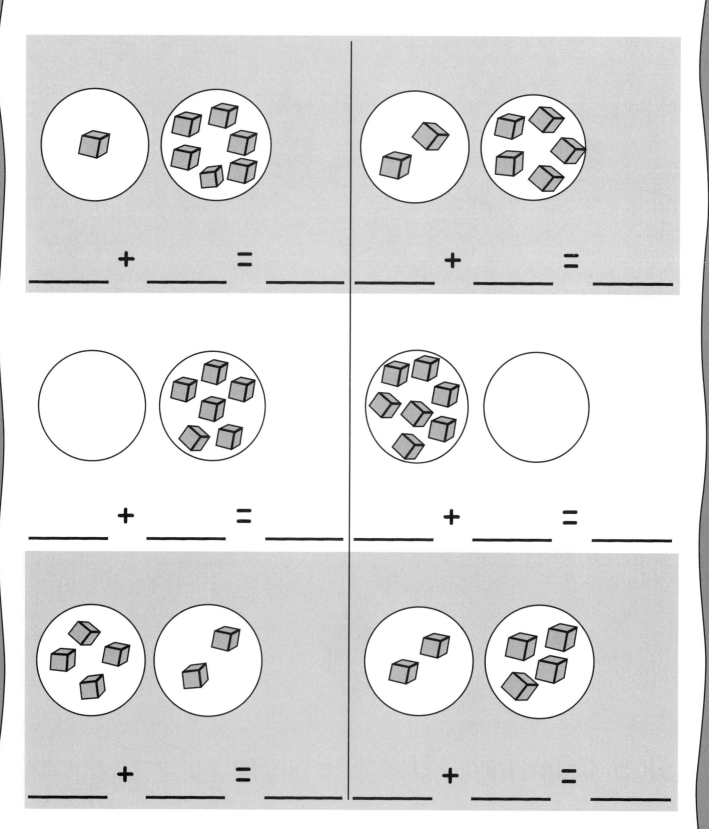

Write each numeral, then say each number sentence.

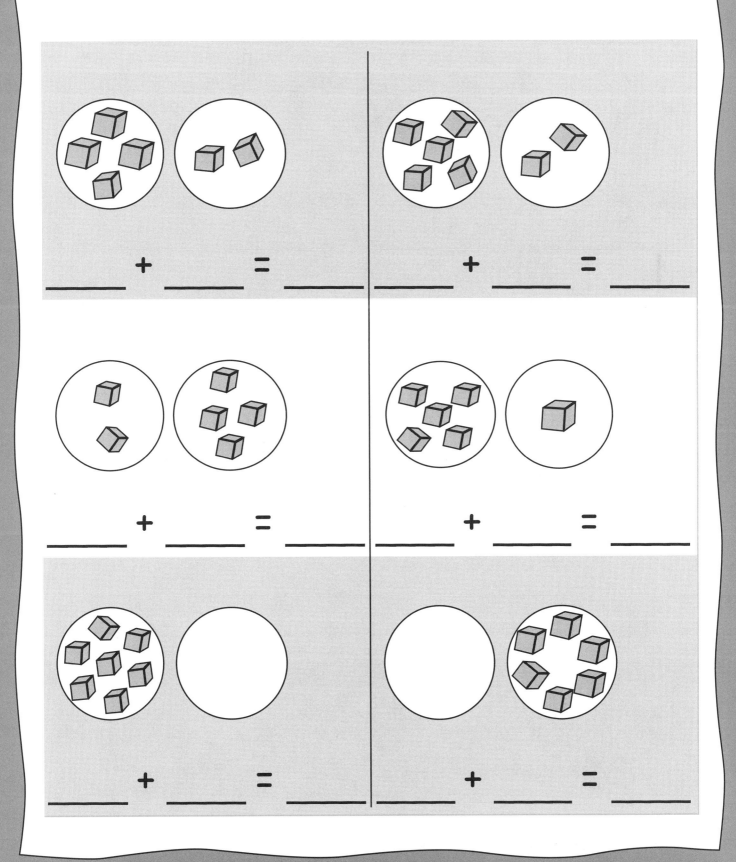

Skip and Count by 2s (to 40)

Begin with 2, then touch and say the number on each gray star.

Trace the dotted numerals on the pattern, then write the numerals that complete the pattern.

1 2   1 2   1 2   _ _ _

0 1 2   0 1 2   _ _ _ _

5 6 7   5 6 7   _ _ _

6 7 8   6 7 8   _ _ _

7 8 9   7 8 9   _ _ _

Trace the dotted numerals on the pattern, then write the numerals that complete the pattern.

3 4 5   3 4 5   ___ ___ ___

2 3 4   2 3 4   ___ ___ ___

4 5 6   4 5 6   ___ ___ ___

6 7 8   6 7 8   ___ ___ ___

7 8 9   7 8 9   ___ ___ ___

0 1 2   0 1 2   ___ ___ ___

Draw, write, or say the answer for each of the following problems.

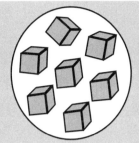 − 2 = _____

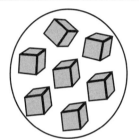 − 5 = _____

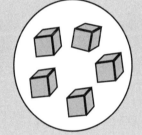 − 5 = _____

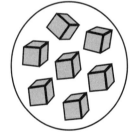 − 3 = _____

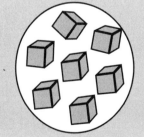 − 0 = _____

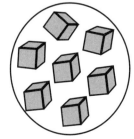 − 1 = _____

Count and Color

Color 7

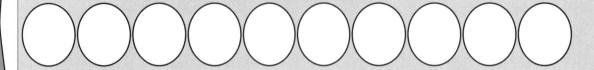

Color 8

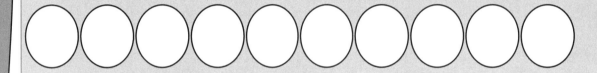

Color 9

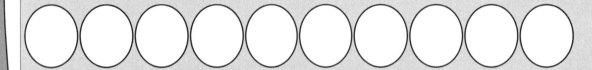

Color 10

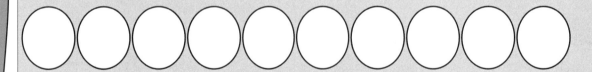

Color 11

Color 12

Are these children going
swimming? How do you know?

Are these girls cold?
How do you know?

Are these boys warm?
How do you know?

Put an X on the
thermometer that shows
the warmer temperature.

Trace each numeral, then draw a line segment to its matching picture.

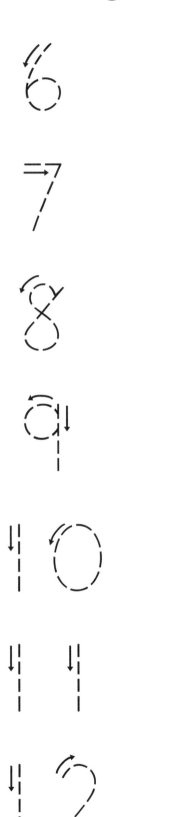

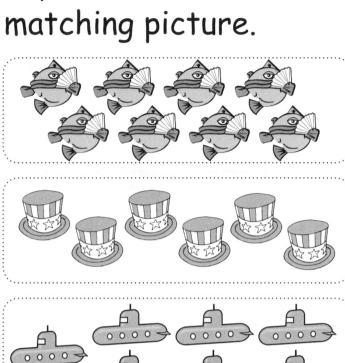

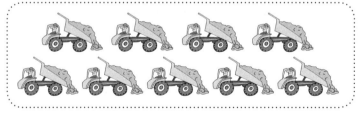

An object that is split into three parts of the same size has three thirds. Circle each object that is split into three parts of the same size (thirds). Put an X on each object that is not.

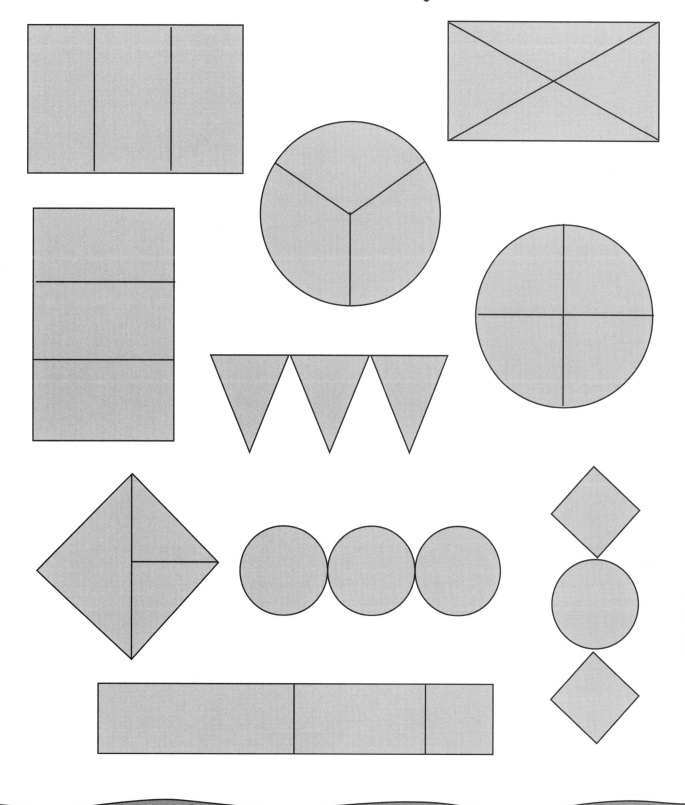

Color $\frac{1}{3}$ of each object.

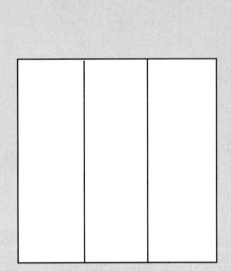 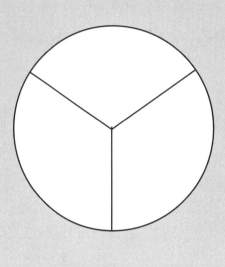

Color $\frac{1}{3}$ of each object.

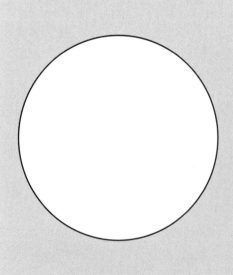  

Use line segments to match each coin to its name and the value.

 penny  25¢

 **nickel** 5¢

 dime 10¢

 quarter 1¢

A store has 4 dogs and 7 cats.
Create a bar graph that shows this.

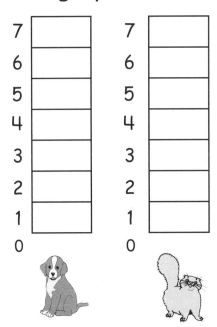

A farmer has 6 pigs and 6 horses. Create a bar graph that shows this.

A box of crayons has 7 black and 6 white crayons. Create a bar graph that shows this.

A child has 6 black socks and 4 white socks. Create a bar graph that shows this.

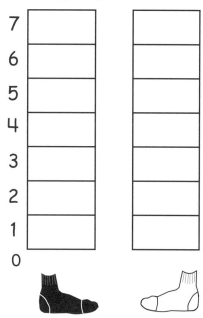

A child has 2 teddy bears and 7 dolls. Create a bar graph that shows this.

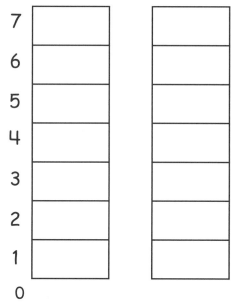

A child has 6 balls and 6 bats. Create a bar graph that shows this.

Trace each numeral, then draw a line segment to its matching picture.

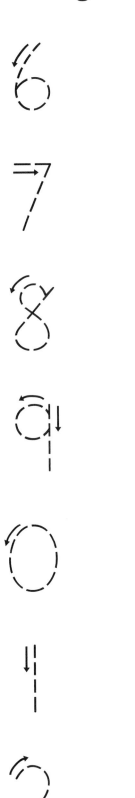

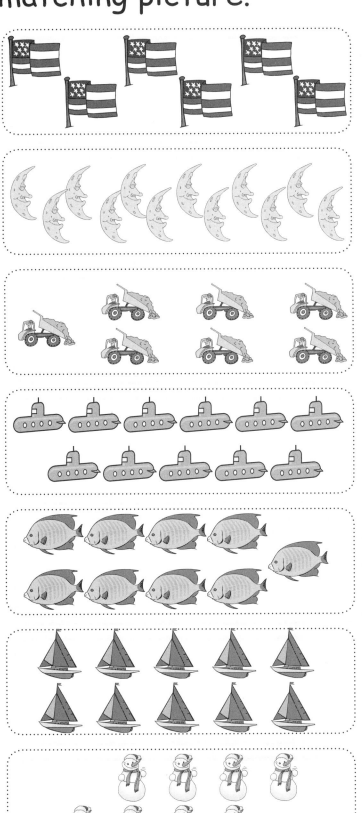

Skip and Count by 2s (to 50)

Touch and say the number on each white ☆ star.

Complete each number sentence to find the sum of fruit on both trees.

 +  =

 +  =

 +  =

 +  = _____

Trace each numeral, then draw a line segment to its matching picture.

 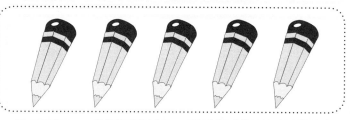

 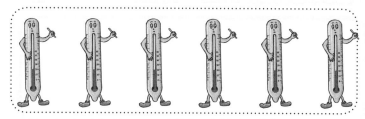

Add the fruit on each pair of trees to complete the number sentence.

 +  = _____

 +  = _____

 +  = _____

 +  = _____

How many black faces? _____
Even or Odd?

How many gray faces? _____
Even or Odd?

How many white faces? _____
Even or Odd?

How many black faces? _____
Even or Odd?

How many gray faces? _____
Even or Odd?

Even or Odd
Draw line segments to connect pairs, then answer the questions.

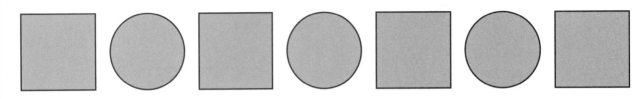

How many pairs? _____ Even or Odd number of shapes?

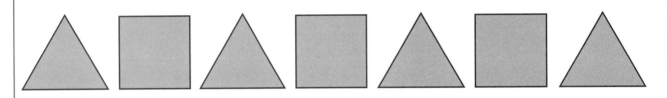

How many pairs? _____ Even or Odd number of shapes?

How many pairs? _____ Even or Odd number of boys?

How many pairs? _____ Even or Odd number of girls?

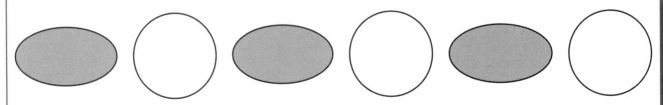

How many pairs? _____ Even or Odd number of shapes?

Count to 10 and Back

Help the bunny find the carrot. Touch and say each number as you count to 10.

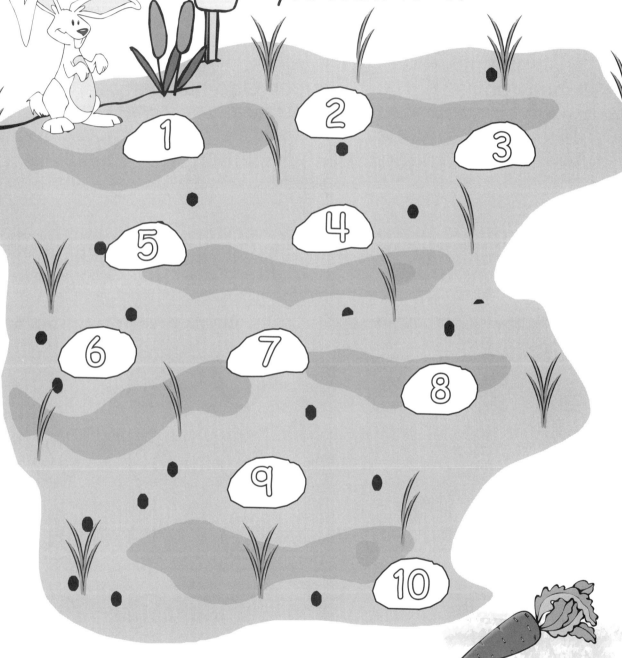

Now, show the bunny how to get back home. Begin with 10. Touch and say each number as you count backwards.

Trace each numeral, then circle the correct answer.

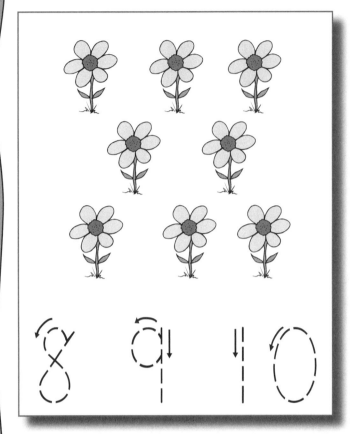

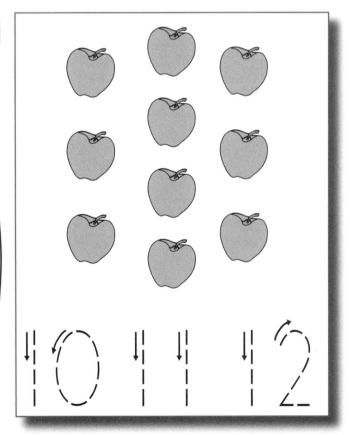

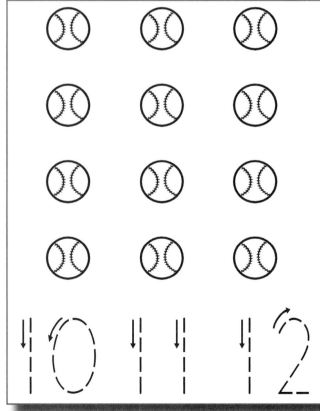

Skip and Count by 5s (to 20)

Begin at △5△. Touch and say the numeral on each white triangle.

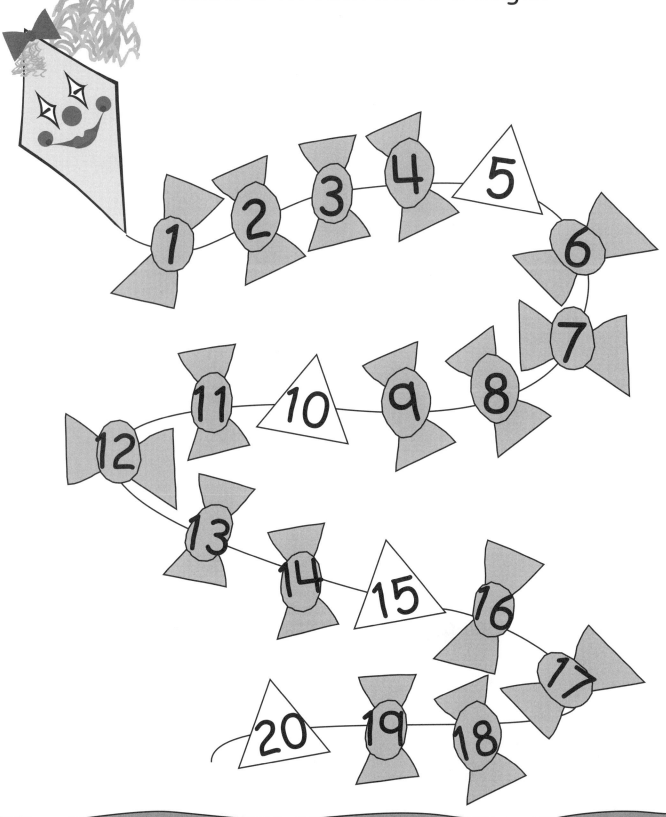

Put a dot • on all of the triangles and squares.

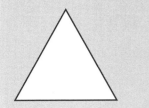    

Put an X on all of the circles and triangles.

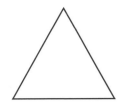 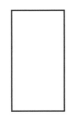 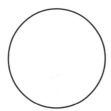 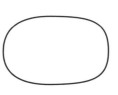 

Put a dot • on all of the shapes without corners.

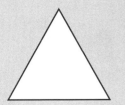    

Put an X on all of the shapes that have corners.

  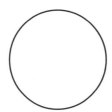 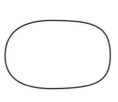 

Put a dot • on all of the rectangles that are not square.

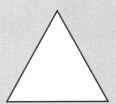    

Put a dot • on the picture of a box.

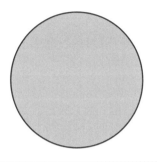 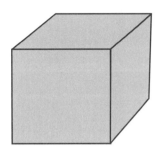 

Put an X on a picture of a ball with just two colors.

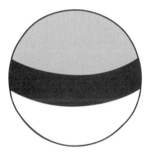  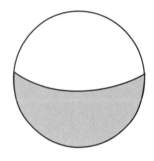

Put a dot • on the picture of a pyramid.

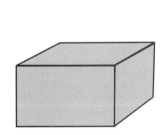 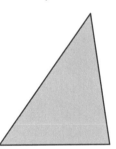 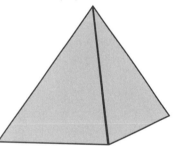 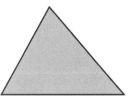

Put an X on the box that probably would be a cereal box.

 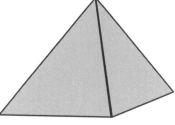  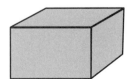

Trace each numeral, then circle the correct amount.

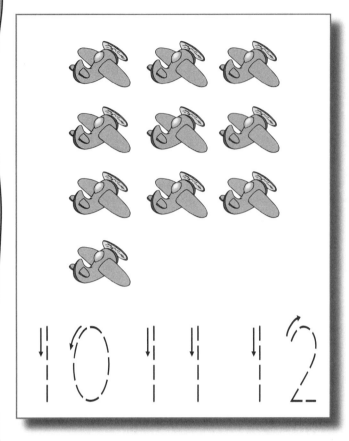

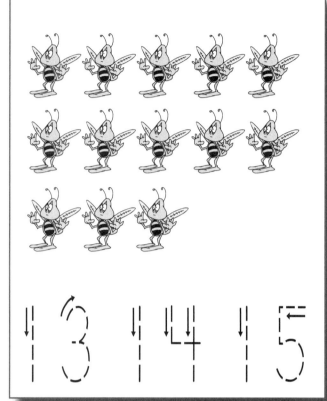

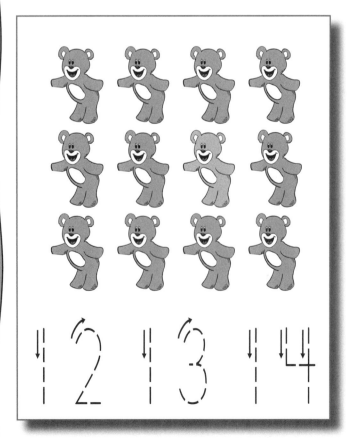

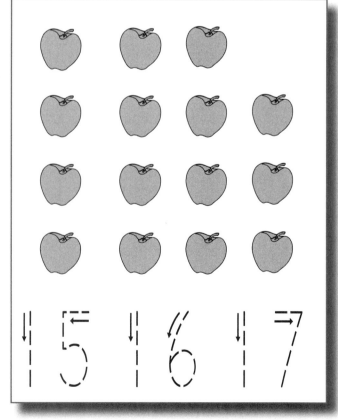

Complete each number sentence, then say each number sentence.

If you pick one orange, how many will be left on the tree?

_____ - _____ = _____

If you pick three apples, how many will be left on the tree?

_____ - _____ = _____

If you pick two oranges, how many will be left on the tree?

_____ - _____ = _____

If you pick four apples, how many will be left on the tree?

_____ - _____ = _____

Help the little duck get back to his family. Draw a path to connect the footprints in the correct order counting by fives.

An object that is split into four parts of the same size has four quarters. Circle each object that is split into four parts of the same size (quarters). Put an X over each object that is not.

Color $\frac{1}{4}$ of each object.

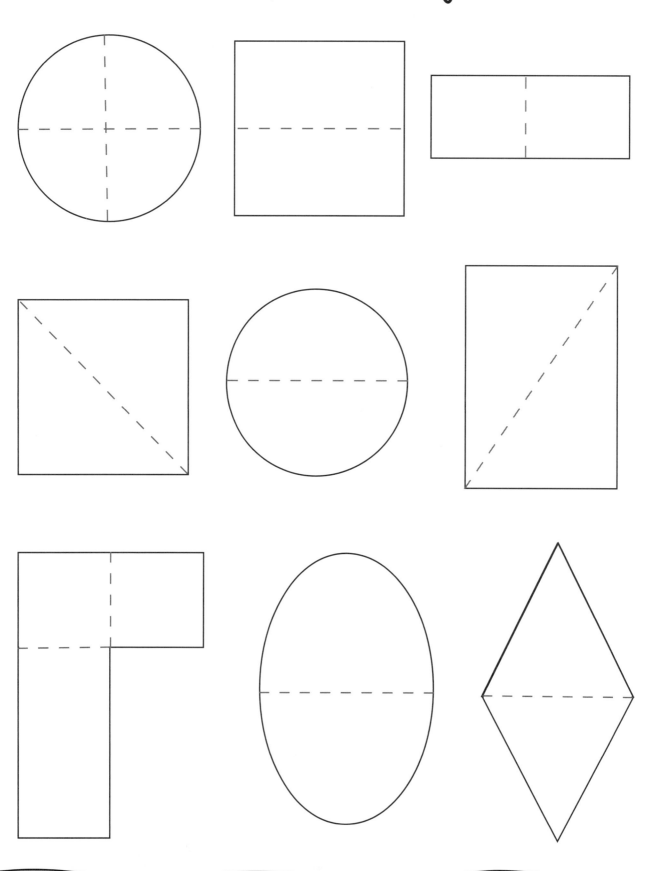

Color one fourth ($\frac{1}{4}$) of each set.

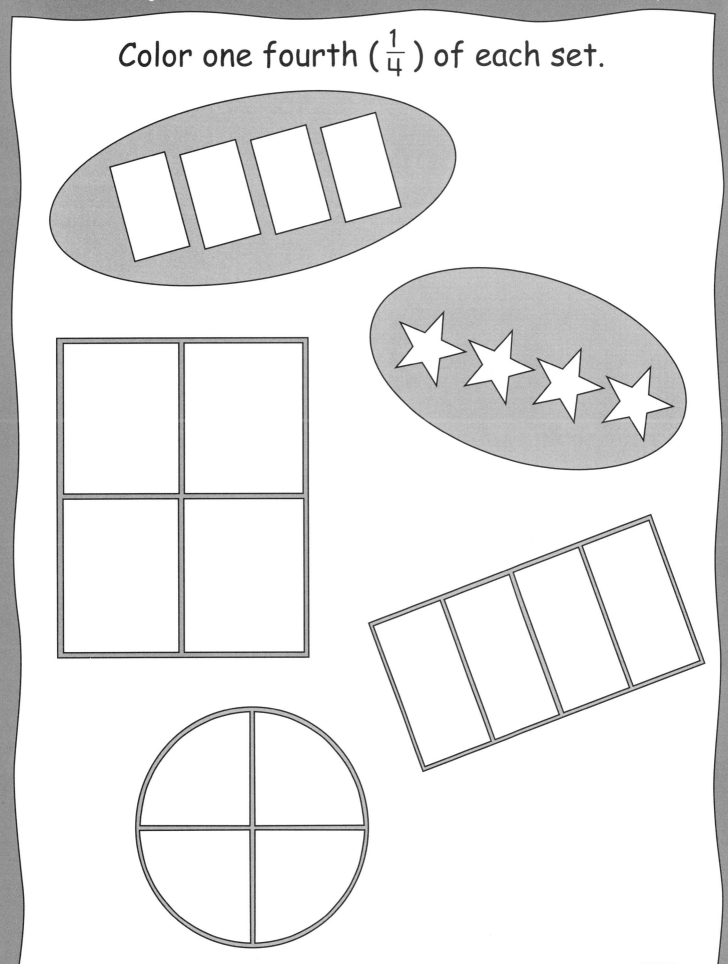

Heavier or Lighter?

1. Which is heavier, your foot or your leg and foot?

2. Which is lighter, you or a baseball?

3. Who is heavier, your mom or you?

4. Which is lighter, a horse or a dog?

5. Which is lighter, a baseball bat or 4 baseball bats?

6. Which is heavier, all of your toys or 3 of your toys?

7. Which is heavier, a car or a bike?

8. Which is heavier, a school bus or a skateboard?

Skip and Count by 5s (to 50)

Begin at △5△, then touch and say the number
on each white triangle.

Heavier or Lighter?

1. Which is heavier, your bicycle or your family car?

2. Which is lighter, you or a baseball?

3. Who is heavier, you or this book?

4. Which is lighter, a baseball bat or you?

5. Which is lighter, a softball or a baseball?

6. Which is heavier, all of your toys, some of your toys, or most of your toys?

7. Which is heavier, 6 marbles all the same size, half of those marbles, or most of those marbles?

8. Which is heavier, 8 marbles all the same size, or all of those marbles plus another marble?

On a number line, 3 + 4 = 7 would appear as below. Use each number line to show its number sentence.

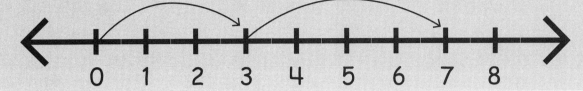

0 1 2 3 4 5 6 7 8

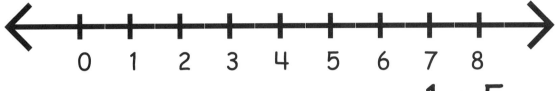

0 1 2 3 4 5 6 7 8

1 + 5 = _____

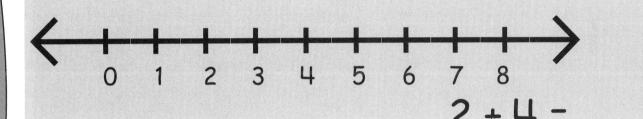

0 1 2 3 4 5 6 7 8

2 + 4 = _____

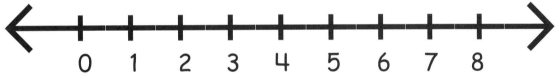

0 1 2 3 4 5 6 7 8

4 + 2 = _____

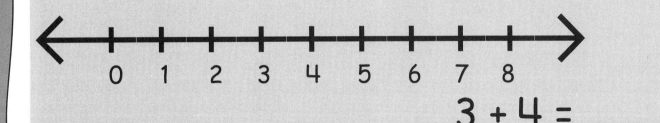

0 1 2 3 4 5 6 7 8

3 + 4 = _____

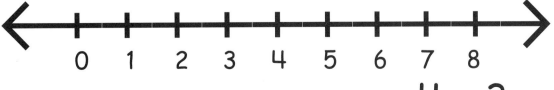

0 1 2 3 4 5 6 7 8

4 + 3 = _____

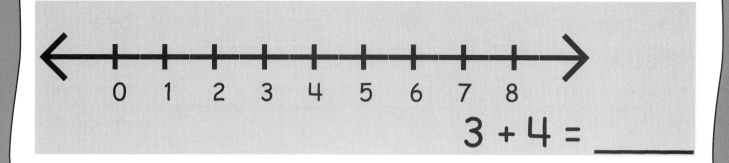

$3 + 4 =$ _____

$4 + 3 =$ _____

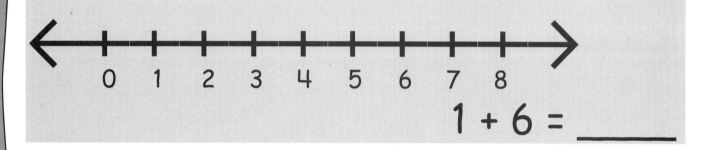

$1 + 6 =$ _____

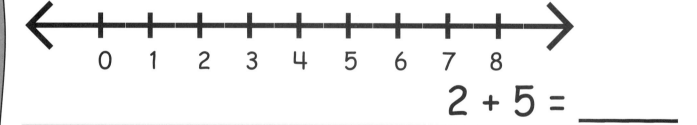

$2 + 5 =$ _____

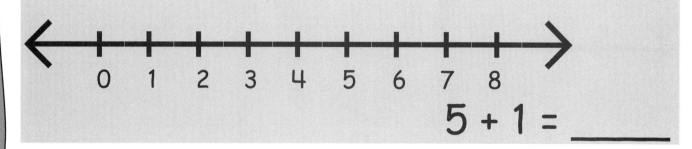

$5 + 1 =$ _____

$5 + 2 =$ _____

A store has 4 dogs and 7 cats.
Create a bar graph that shows this.

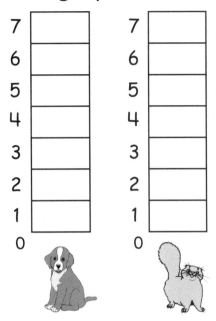

A baseball player has 2 gloves, 7 balls, and 3 bats. Create a bar graph that shows this.

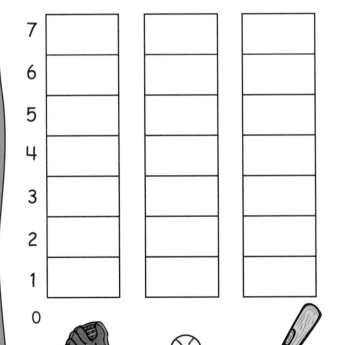

A family has 1 car, 1 mini-van, 7 bikes, and 5 skateboards. Create a bar graph that shows this.

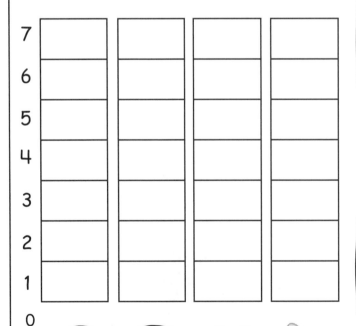

Touch and say the numerals in the white rectangles. Begin with 10.

1	2	3	4	5	6	7	8	9	**10**
11	12	13	14	15	16	17	18	19	**20**
21	22	23	24	25	26	27	28	29	**30**
31	32	33	34	35	36	37	38	39	**40**
41	42	43	44	45	46	47	48	49	**50**

A family has 2 cars, 7 bicycles, and 4 skateboards.
Create a bar graph that shows this.

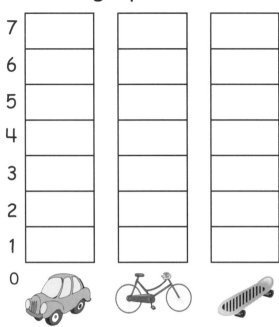

A child has 1 bicycle, 1 skateboard, 1 pair of skates, and 6 helmets. Create a bar graph that shows this.

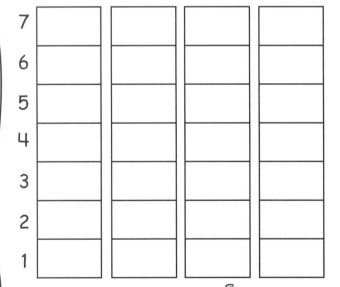

A toy box holds 7 balls, 3 bats, 5 gloves, and 4 helmets. Create a bar graph that shows this.

| 7 |
| 6 |
| 5 |
| 4 |
| 3 |
| 2 |
| 1 |
| 0 |

Count the tally marks in each set and say or write the sum of the tally marks.

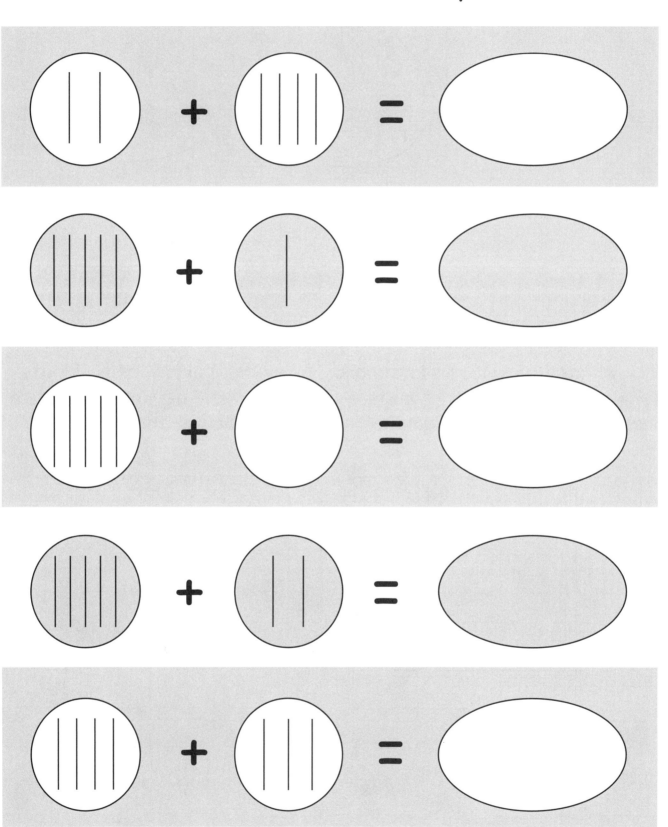

Count the tally marks in each set and say or write the sum of the tally marks.

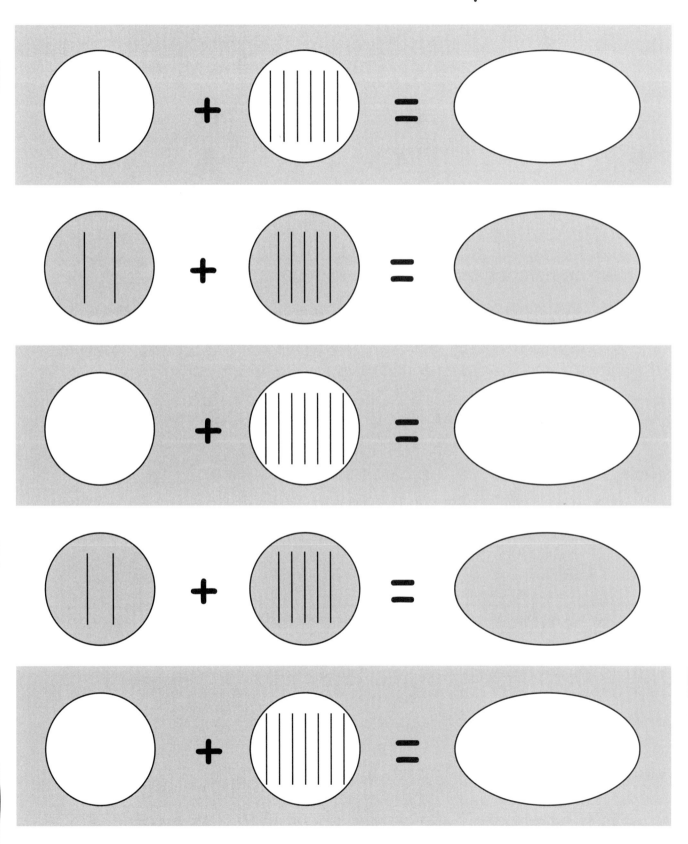

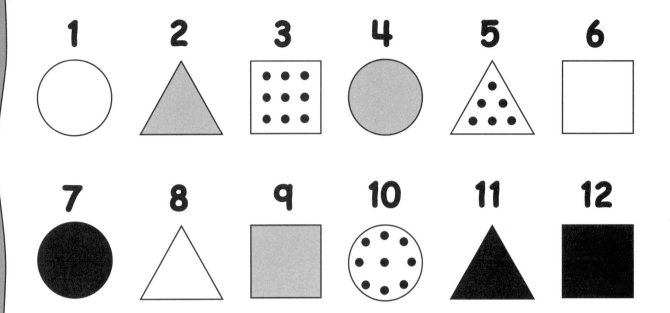

What is the number of the gray and square shape? _____

What is the number of the round and dotted shape? _____

What is the number of the white and square shape? _____

What is the number of the gray and triangle shape? _____

What is the number of the black and round shape? _____

What is the number of the dotted and square shape? _____

What is the number of the black and triangle shape? _____

What is the number of the white and triangle shape? _____

What is the number of the round and gray shape? _____

What is the number of the black shape with 4 corners? _____

What is the number of the white circle? _____

What is the number of the dotted shape with just 3 corners? _____

Write the answer to each problem in the space provided.

1 + 5 = ___	2 + 4 = ___
3 + 4 = ___	0 + 6 = ___
2 + 3 = ___	3 + 2 = ___
1 + 6 = ___	5 + 2 = ___
7 + 0 = ___	5 + 1 = ___
1 + 4 = ___	2 + 5 = ___
4 + 2 = ___	3 + 3 = ___
5 + 1 = ___	6 + 0 = ___
1 + 5 = ___	2 + 4 = ___
5 + 2 = ___	0 + 0 = ___

Circle the objects that could be made with just these four shapes. Each shape can only be used once. Explain each of your answers.

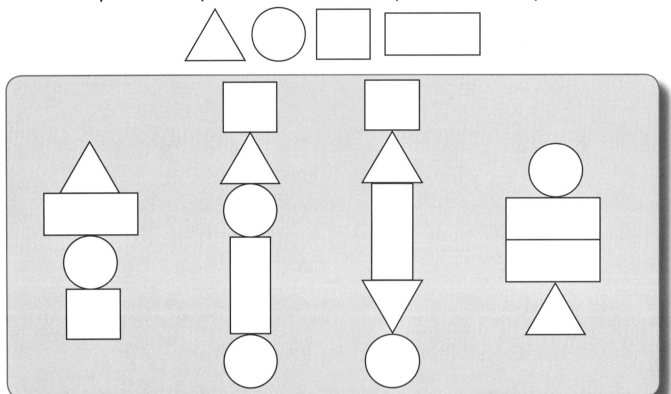

Circle the objects that could be made with just these four shapes. Each shape can only be used once. Explain each of your answers.

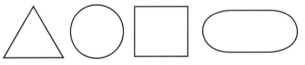

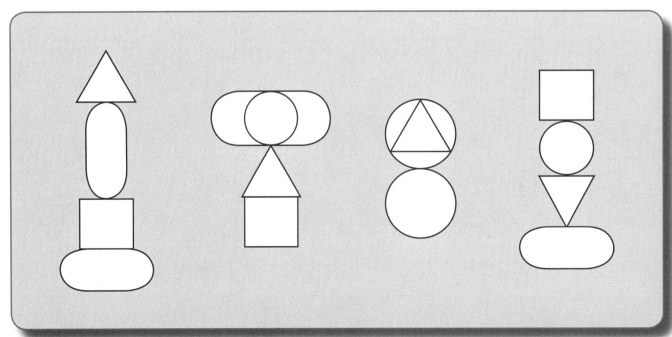

Starting at the left, color the first and second circles red. Color the last two circles blue.

How many circles are not colored? _____

Starting at the left, color the first and second circles blue. Color the 3rd and 4th squares green.

How many circles are not colored? _____

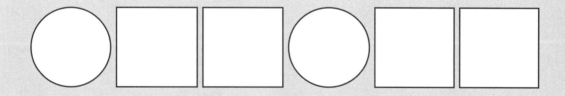

Starting at the left, color the second triangle green. Color the first circle blue. Color the fourth triangle red.

How many triangles are not colored? _____

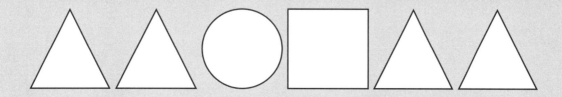

Help the mother hen find her eggs. Draw a
path to connect the footprints in the correct
order, counting by tens.

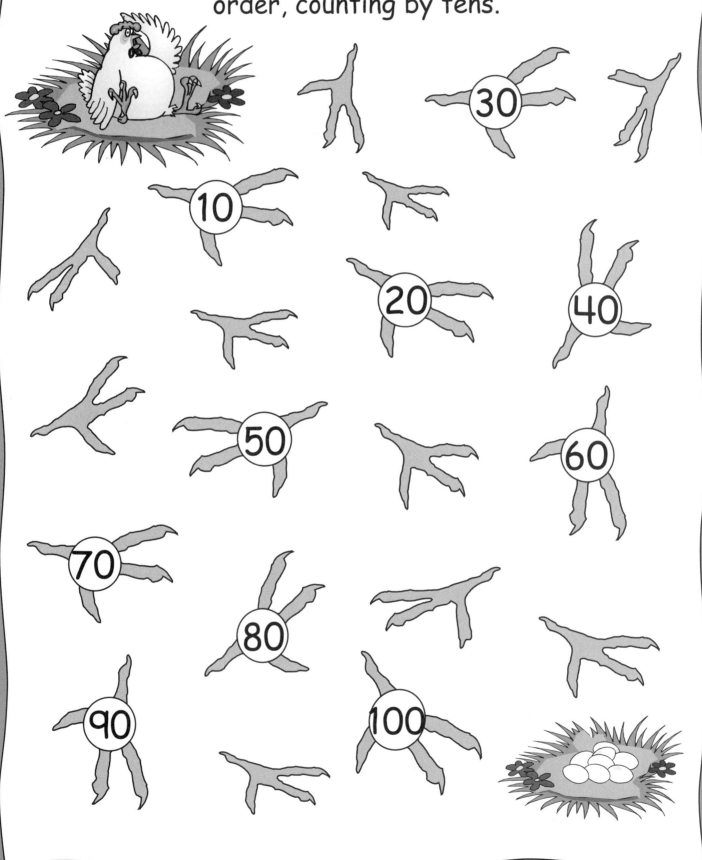

Write the answer to each problem in the space provided.

1 + 4 = ___	2 + 5 = ___
4 + 2 = ___	3 + 3 = ___
3 + 4 = ___	0 + 6 = ___
1 + 6 = ___	5 + 2 = ___
1 + 5 = ___	2 + 4 = ___
2 + 3 = ___	3 + 2 = ___
7 + 0 = ___	5 + 1 = ___
5 + 1 = ___	6 + 0 = ___
1 + 5 = ___	2 + 4 = ___
5 + 2 = ___	0 + 0 = ___

Use line segments to match each coin to its name and the value.

penny

nickel

dime

quarter

Counting by tens, use line segments to connect the stones that a little bunny will hop on to get a carrot. Then color the stones.

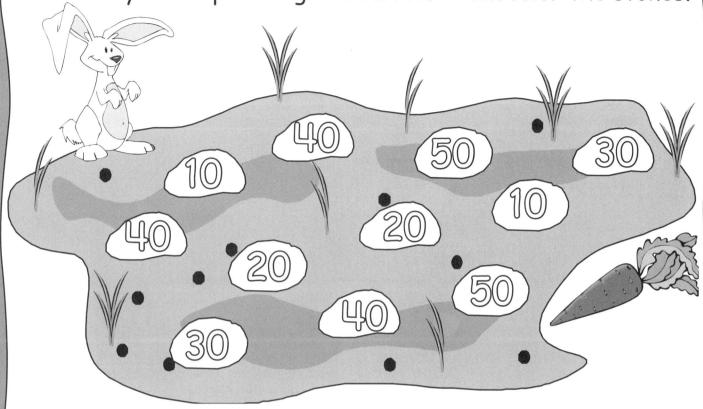

Counting by tens, use line segments to connect the lily pads that Froggy will hop on to cross the stream. Then color the pads.

Complete the number sentence to each problem.

There are 3 books on one shelf and 3 more books on another shelf. How many books are there in all?

_____ + _____ = _____

You ate 3 grapes, then you ate 4 more grapes. How many grapes did you eat?

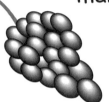

_____ + _____ = _____

You have your bike. Six friends bring their bikes to your house, too. How many bikes are at your house?

_____ + _____ = _____

You have 2 pennies. You find 3 pennies. Later you find another penny. How many pennies do you have?

_____ + _____ + _____ = _____

There are 6 white beads on a string. You add 2 black beads to the string. How many beads are on the string?

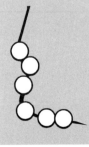

_____ + _____ = _____

Venn Diagrams

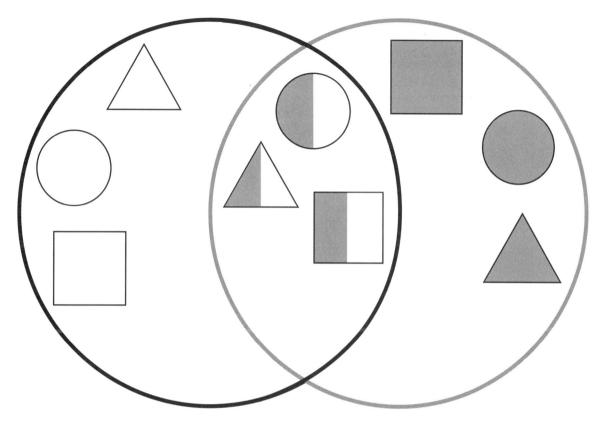

1. Describe the objects that are in both the black and gray circles.

2. What is the same about every object in the black circle?

3. What is the same about every object in the gray circle?

4. Point to where each object belongs in the picture.

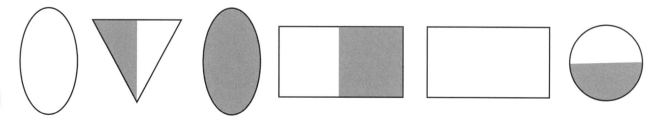

Color some circles red but most circles green.

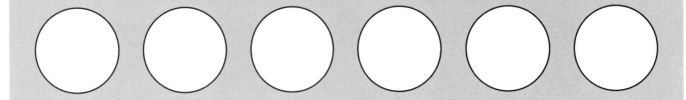

Color most circles red but some circles green.

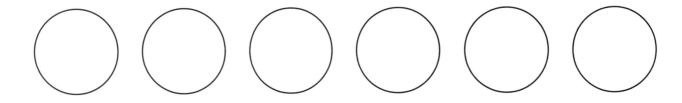

Color half the circles red.

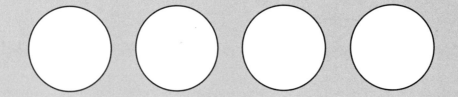

Color an even number of circles.

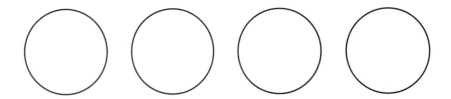

Color an odd number of circles.

Complete the number
sentence to each problem.

There are 2 books on one shelf and 3 more books on another shelf. How many books are there in all?

_____ **+** _____ **=** _____

You ate 2 grapes, then you ate 6 more grapes. How many grapes did you eat?

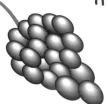

_____ **+** _____ **=** _____

You have your bike. Five friends bring their bikes to your house, too. How many bikes are at your house?

_____ **+** _____ **=** _____

You have 2 pennies. You find 3 pennies. Later you find 2 more pennies. How many pennies do you have?

_____ **+** _____ **+** _____ **=** _____

There are 6 white beads on a string. You add 2 black beads to the string. Then you add 1 gray bead to the string. How many beads are on the string?

_____ **+** _____ **+** _____ **=** _____

Trace the numerals, then write the numerals that continue the pattern.

1 2 1 2 1 2 ___ ___

2 4 2 4 2 4 ___ ___

3 5 3 5 3 5 ___ ___

5 7 5 7 5 7 ___ ___

7 2 7 2 7 2 ___ ___

6 5 6 5 6 5 ___ ___

1 2 3 1 2 3 ___ ___ ___

2 4 8 2 4 8 ___ ___ ___

1 4 9 1 4 9 ___ ___ ___

When you divide something into two parts and both halves look exactly the same, we say the object is **symmetric**, or has **symmetry**.

For example, this square is divided by the dotted line into two parts that are exactly the same. The line that does the dividing is called the **line of symmetry**.

Circle all the symmetric objects below. Put an X over all the objects that are not symmetric.

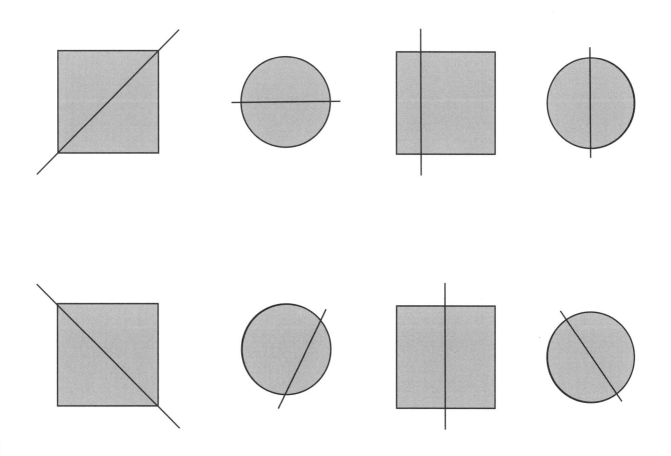

Circle all the symmetric objects below. Put an X over all the objects that are not symmetric.

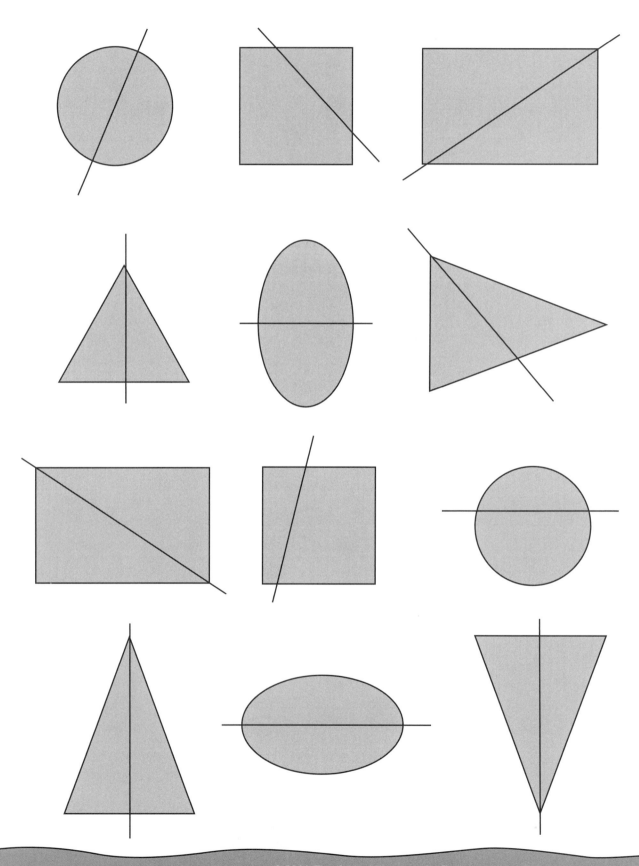

How many of each animal is in the pet store?

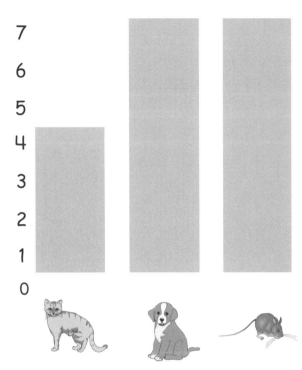

How many sodas did each child drink?

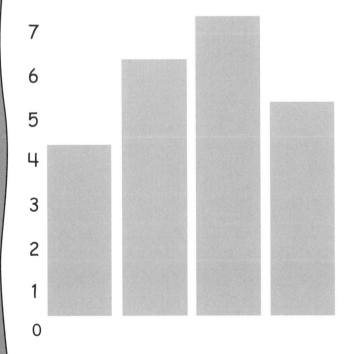

How many home runs did each player hit?

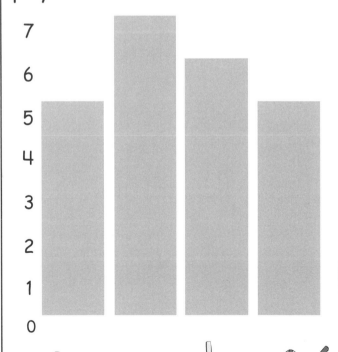

Draw a line of symmetry to split each object in half.

Put an X on the glass that would hold more if they are both full.

Put an X on the glass that would hold less if they are both full.

Put an X on the glass that would hold less if they are all full.

Put an X on the glass that would hold more if they are all full.

Put an X on the glass that would not hold the most or the least if they are all full.

Help the little bear cross the river. Use line segments to draw a path from 10 to 100. Count by tens.

Count the fruit on each tree, then complete the number sentence.

   − 3 = ___

   − 3 = ___

   − 2 = ___

   − 1 = ___

Complete each number sentence, then say each number sentence.

If you took one banana from the plate, how many bananas would there be?

_____ - _____ = _____

If you took three cherries from the bowl, how many cherries would there be?

_____ - _____ = _____

If you took three pears from the plate, how many pears would there be?

_____ - _____ = _____

If you took three apples from the bowl, how many apples would there be?

_____ - _____ = _____

Complete each number sentence, then say each number sentence.

If you pick four oranges, how many will be left on the tree?

_____ - _____ = _____

If you pick five oranges, how many will be left on the tree?

_____ - _____ = _____

If you pick six oranges, how many will be left on the tree?

_____ - _____ = _____

If you pick seven oranges, how many will be left on the tree?

_____ - _____ = _____

Count the tally marks in each set and say or write how many tally marks are left.

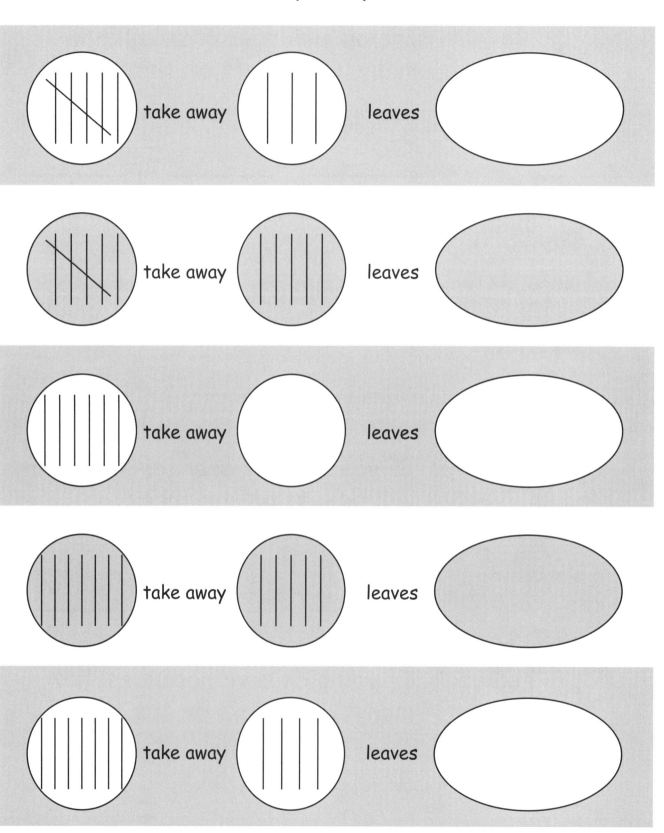

take away   leaves

take away   leaves

take away   leaves

take away   leaves

take away   leaves

Count the tally marks in each set and say or write how many tally marks are left.

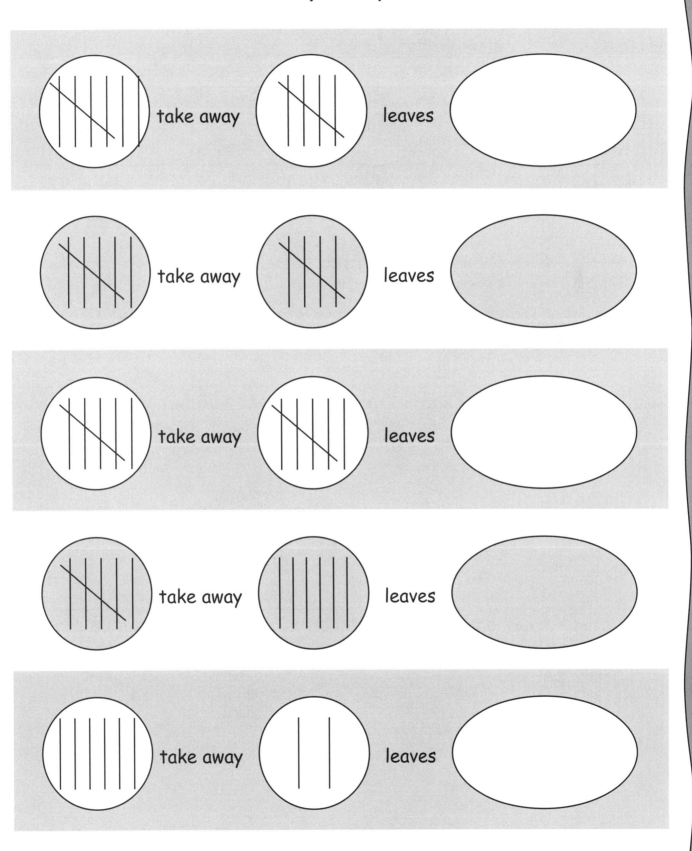

You have 6 gray balls and 1 black ball. If you put them in a box, closed your eyes, and picked only one ball, what color would it probably be?

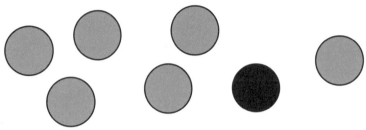

You have 7 gray balls. If you put them in a box, closed your eyes, and picked only one ball, what color would it be?

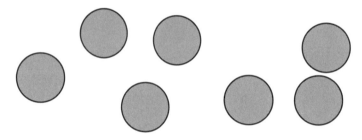

You have 4 white balls and 2 black balls. If you put them in a box, closed your eyes, and picked only one ball, what color would it probably be?

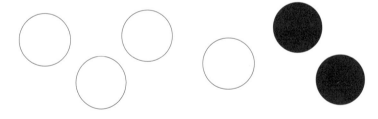

You have 4 gray balls and 3 white balls. If you put them in a box, closed your eyes, and picked only one ball, what color would it probably be?

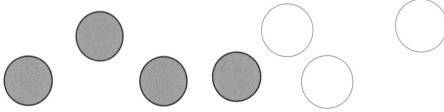

Use the number line to show each number sentence.

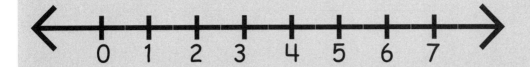

6 - 2 = _____

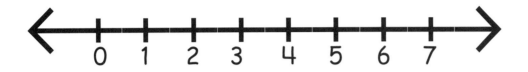

7 - 3 = _____

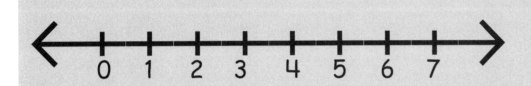

6 - 4 = _____

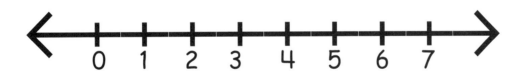

7 - 2 = _____

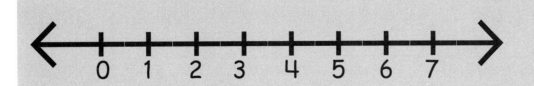

7 - 1 = _____

Write or Say

If you put these socks in a box, closed your eyes, and picked only one sock, what color would it probably be, and why?

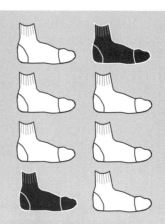

If you put these marbles in a box, closed your eyes, and picked only one marble, what color would it probably be, and why?

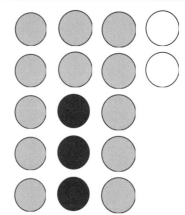

If you put these socks in a box, closed your eyes, and picked only one sock, what color would it probably be, and why?

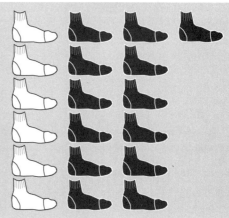

If you put these books in a box, closed your eyes, and picked only one book, what would it probably be, and why?

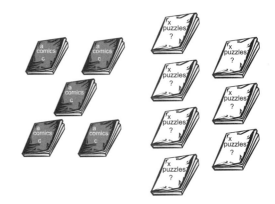

Use the number line to show each number sentence.

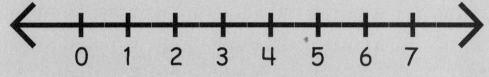

6 - 1= _____

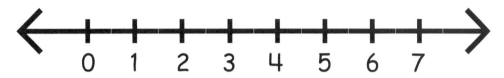

7 - 5= _____

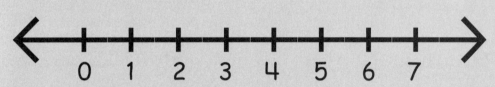

6 - 5= _____

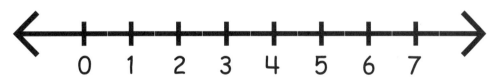

6 - 2= _____

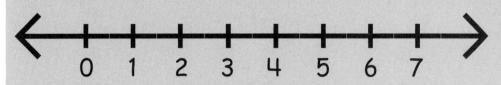

7 - 2= _____

Write the difference to each subtraction problem.

7 - 2 = ___	5 - 2 = ___
6 - 2 = ___	7 - 1 = ___
6 - 3 = ___	7 - 4 = ___
7 - 3 = ___	7 - 6 = ___
7 - 0 = ___	6 - 5 = ___
7 - 1 = ___	7 - 5 = ___
6 - 1 = ___	6 - 3 = ___
6 - 5 = ___	7 - 6 = ___
5 - 4 = ___	6 - 2 = ___
6 - 3 = ___	7 - 0 = ___

Put an X over all objects that are round and black. Circle the objects that have less than four corners.

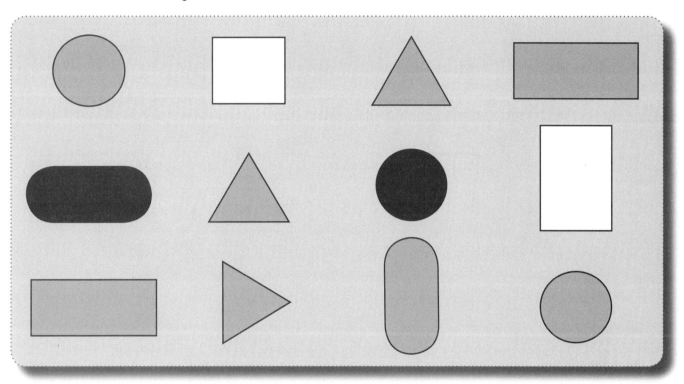

Put a big dot ● on all objects that have more than three corners. Circle the objects without dots ● that are black or white.

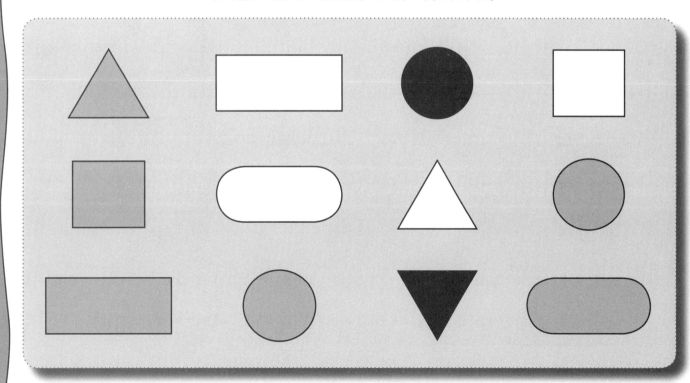

Draw a line segment to show where each piece was removed (cut).

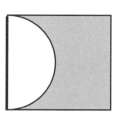

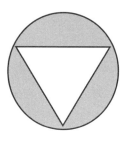

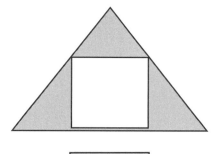

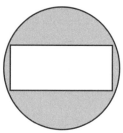

Create a bar graph that shows more mice than birds.

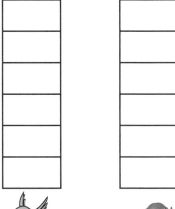

Create a bar graph that shows four men, fewer boys than men, and fewer girls than men.

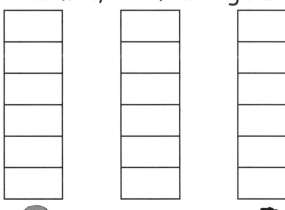

Create a bar graph that shows three dogs, more cats than dogs, and more fish than dogs.

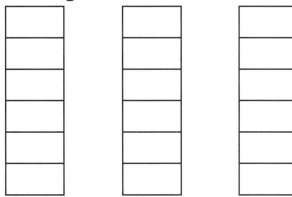

Write the difference to each subtraction problem.

7 - 1 = ___	7 - 5 = ___
6 - 3 = ___	7 - 4 = ___
7 - 3 = ___	7 - 6 = ___
7 - 2 = ___	5 - 2 = ___
6 - 2 = ___	7 - 1 = ___
6 - 1 = ___	6 - 3 = ___
5 - 4 = ___	6 - 2 = ___
6 - 5 = ___	7 - 6 = ___
7 - 0 = ___	6 - 5 = ___
6 - 3 = ___	7 - 0 = ___

One Year

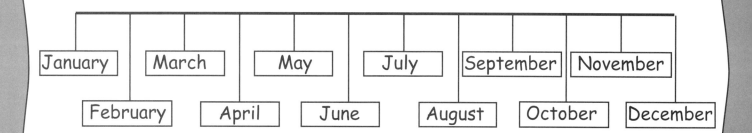

1. Point to and say the names of two summer months.

2. How many months in a year? _____

3. Point to and say the name of the first month in the year.

4. Point to and say the name of the last month of the year.

5. Point to and say the name of the two months in the middle of the year.

6. Point to and say the name of the second month of the year.

7. Point to and say the name of the third month of the year.

8. Point to and say the name of the fifth month of the year.

Complete the number
sentence for each problem.

You have 7 toys and give 3 to a friend. How many toys do you have after that?

_____ - _____ = _____

There are 7 books on a shelf in your room. You return 2 to the library. How many books are left on the shelf?

_____ - _____ = _____

You took 6 swings at balls on a batting T. You got 4 hits. How many times did you miss the ball on the batting T?

_____ - _____ = _____

You have 7 pennies. You spend 5 pennies. How many pennies do you have left?

_____ - _____ = _____

There were 7 white beads on a string. Then you took 7 white beads off. How many beads were left on the string?

_____ - _____ = _____

Complete each number sentence, then say each number sentence.

If you pick three oranges, how many will be left on the tree?

_____ - _____ = _____

If you pick five oranges, how many will be left on the tree?

_____ - _____ = _____

If you pick two oranges, how many will be left on the tree?

_____ - _____ = _____

If you pick seven oranges, how many will be left on the tree?

_____ - _____ = _____

Complete the number sentence for each problem.

You have 6 toys and give 5 to a friend. How many toys do you have after that?

_____ - _____ = _____

There are 7 books on a shelf in your room. You return 6 to the library. How many books are left on the shelf?

_____ - _____ = _____

You took 7 swings at balls on a batting T. You got 2 hits. How many times did you miss the ball on the batting T?

_____ - _____ = _____

You have 5 pennies. You spend 5 pennies. How many pennies do you have left?

_____ - _____ = _____

There are 7 white beads on a string. You take off 2 beads. How many beads are left?

_____ - _____ = _____

Put an X on the box that would hold more if both boxes were full.

Put an X on the box that would hold less if both boxes were full.

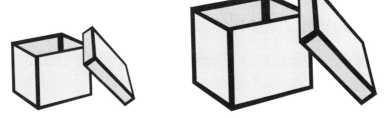

Put an X on the box that would hold less if all the boxes were full.

Put an X on the box that would hold more if all the boxes were full.

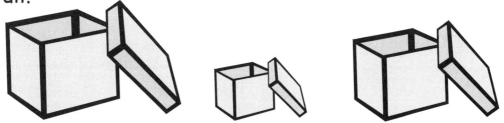

Put an X on the box that would not hold the most or the least if all the boxes were full.

Locate 11 on the number line by tracing the path from 0 to 11.

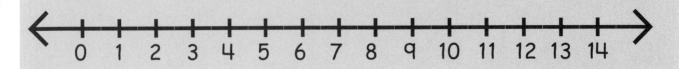

Locate 14 on the number line by tracing the path from 0 to 14.

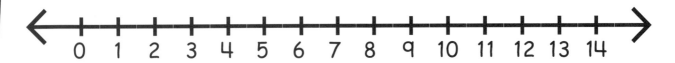

Locate 13 on the number line by tracing the path from 0 to 13.

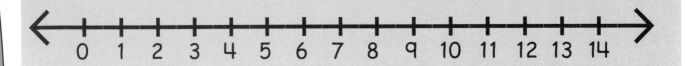

Locate 12 on the number line by tracing the path from 0 to 12.

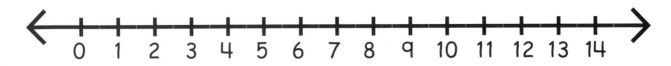

Locate 10 on the number line by tracing the path from 0 to 10.

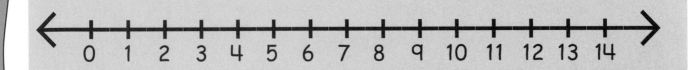

The Fruit Stand

Each giant strawberry costs a nickel. Count the strawberries by 5s. How much are all the strawberries? _____ ¢

Each slice of watermelon costs a nickel. Count the slices by 5s. How much are all the watermelon slices? _____ ¢

Locate 8 on the number line by tracing the path from 0 to 8.

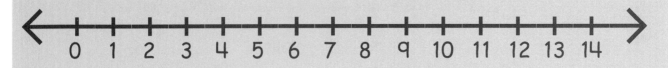

Locate 10 on the number line by tracing the path from 0 to 10.

Locate 13 on the number line by tracing the path from 0 to 13.

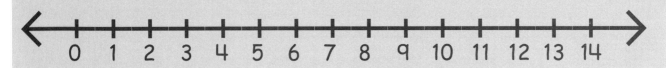

Locate 11 on the number line by tracing the path from 0 to 11.

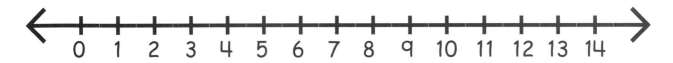

Locate 9 on the number line by tracing the path from 0 to 9.

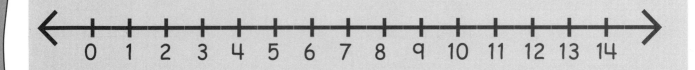

Trace the numerals, then write the numerals that continue the pattern.

1 2 1 2 1 2 ___ ___

2 4 2 4 2 4 ___ ___

3 5 3 5 3 5 ___ ___

5 7 5 7 5 7 ___ ___

7 2 7 2 7 2 ___ ___

1 2 3 1 2 3 ___ ___ ___

Trace the numerals, then write the numerals that continue the pattern.

6 5 6 5 6 5 __ __

1 2 3 1 2 3 __ __ __

2 4 8 2 4 8 __ __ __

1 4 9 1 4 9 __ __ __

1 1 2 1 1 2 __ __ __

3 1 4 3 1 4 __ __ __